JTG Standards and Specifications

This series includes the standards and specifications of the highway transportation industry, issued by the Ministry of Transport of the People's Republic of China. It covers the standards and specifications in terms of technology, administration and service for the process from highway planning through to highway maintenance. The criteria for safety, environment and economic assessment are also included.

Wanping Wu

Specifications for Design of Highway Substructure

人民交通出版社
China Communications Press

Springer

Wanping Wu
CCCC Second Highway
Consultants Co., Ltd.
Beijing, China

ISSN 3004-8818 ISSN 3004-8826 (electronic)
JTG Standards and Specifications
ISBN 978-981-96-2312-9 ISBN 978-981-96-2313-6 (eBook)
https://doi.org/10.1007/978-981-96-2313-6

Jointly published with China Communications Press Co., Ltd.
The print edition is not for sale in China (Mainland). Customers from China (Mainland) please order the print book from: China Communications Press Co., Ltd.
ISBN of the Co-Publisher's edition: 978-7-114-17250-2

This Springer imprint is published by the registered company Springer Nature Singapore Pte Ltd.
The registered company address is: 152 Beach Road, #21-01/04 Gateway East, Singapore 189721, Singapore

Foreword to Chinese Version

According to the requirements of the ministerial circulation paper (2010) No. 132 *Notice on the Preparation and Revision of the Work Plan for the* 2010 *Annual Highway Engineering Standards and Quotas* issued by the Ministry of Transport, CCCC Second Highway Consultants Co., Ltd., has been in charge of the updating of *Specifications for Design of Highway Substructure* (JTG D30—2004).

The new specification is a comprehensive updating of the original *Specifications for Design of Highway Substructure* (JTG D30—2004) (hereinafter referred to as 'the original *Specifications*'). After approval, it will be promulgated and implemented with the *Specifications for Design of Highway Substructure* (JTG D30—2015) (hereinafter referred to as 'the *Specifications*'). The guiding ideology and principles of the updating of the specification summarized the experience and scientific and technological achievements of highway construction projects in China for many years and drawing on the advanced technical methods of relevant standards and norms at home and abroad. The updating is in accordance with the design concept of 'safety, durability, resource conservation, and environmental harmony', fully considers the functional requirements of highway substructure, strengthens the coordinated design of substructure and pavement, and improves the overall strength, stiffness, water stability, temperature stability and durability of the substructure. Furthermore, the updating mainly revised the contents of substructure disease prevention and substructure disease control measures and strives to make the *Specifications* advanced in technology, reasonable in indicators and excellent in operability.

The main technical contents of the *Specifications* consist of seven chapters and 10 appendices, including Chap. 1 General Provisions, Chap. 2 Terms and Symbols, Chap. 3 General Substructure, Chap. 4 Substructure Drainage, Chap. 5 Substructure Protection and Retaining Structures, Chap. 6 Substructure Widening and Reconstruction, and Chap. 7 Special Substructure. The *Specifications* cover all substructure engineering projects involved in highway construction and reconstruction and extension projects.

The updating includes the following main contents:

1. Section 3.2 Roadbed. According to the traffic load level, the range of the roadbed has been adjusted; roadbed design indicators, roadbed rebound modulus control standards and estimation methods of indicators have been added; roadbed treatment measures have been added.
2. Section 3.3 Fill embankment. New provisions have been added in terms of the design principles and methods for determining the height of the embankment; the embankment stability analysis method in Section 3.6 has been modified; the stability safety factor of the embankment and steep slope embankment under rainfall conditions have been added.
3. 'Fly ash embankment' was changed to 'Light weight material embankment' in Sect. 3.9; 'geoplastic foam embankment' and 'foam light soil embankment' were added; the requirements of structural design of light weight material embankment, material design and stability, and checking of settlement were defined.
4. Section 3.10 Embankment with industrial waste material has been added. The technical requirements have been added. Applicable conditions and material requirements for filling embankment such as blast furnace slag, steel slag, and coal gangue have been added; the technical requirements of embankment structure design and embankment stability checking calculation have been added.
5. Chapter 4 Substructure Drainage. The technical requirements for substructure drainage have been added, including the maximum allowable flow rate of open channel, low embankment drainage, digging channel drainage, substructure drainage in the interchange area, central separation zone drainage, seepage well, drainage tunnel and so on.
6. Chapter 5 Substructure Protection and Retaining Structures. New applicable conditions, structural design and material technical requirements for flexible protective structures have been added, including 'geogrid reverse-packaged reinforced earth retaining wall' and 'stone cage retaining wall'; prestressed anchor structure calculation and anti-corrosion requirements, soil nail application conditions, prestressed anchor cable anti-slide pile design requirements and field test and monitoring design requirements have been revised.
7. Chapter 6 Substructure Widening and Reconstruction. Evaluation content of the existing substructure in the expansive soil area and karst area has been added; the existing substructure field test requirements broadened the substructure soft soil foundation treatment measures, and the existing substructure utilization and treatment technical requirements have been revised.
8. Chapter 7 Special Substructure. The design principles, disease prevention measures and technical requirements of 17 types of special substructure have been revised, including landslide, collapse, rock pile, debris flow, karst, soft soil, red clay and high liquid limit soil, expansive soil, loess, saline soil, permafrost, weathering, snow damage, drift ice, goaf, coastal, and reservoir.

9. Section 7.19 Substructure in seasonally frozen soil area has been added. The classification of seasonal frozen soil, the calculation method, and control standard of frost heave, the technical requirements of filling material for substructure and the design requirements of drainage have been proposed.

In the *Specifications*, Chap. 1, Chap. 2 and Chap. 7 (including Section 1, Section 5, Section 6, Section 17, and Section 18) were drafted by Wu Wanping. Section 1, Section 3, Section 4, Section 5, and Section 11 of Chap. 3 were drafted by Mei Shiran and Wu Wanping. Section 2 of Chap. 3 was drafted by Ling Jianming, Lin Xiaoping and Wu Wanping. Section 6 of Chap. 3 was drafted by Deng Weidong. Section 7 of Chap. 3 and Section 5 of Chap. 5 were drafted by Tang Shuming. Section 8 of Chap. 3 were drafted by Sha Aimin. Section 9 of Chap. 3 and Section 6 and Section 7 of Chap. 5 were drafted by Yuan Xizhong. Section 10 of Chap. 3 was drafted by Sha Aimin and Yuan Xizhong. Chapter 4 was drafted by Cheng Ping and Ruan Yanbin. Section 1, Section 2, Section 3 and Section 4 of Chap. 5 were drafted by Zhang Jiaxiang and Wu Wanping. Chapter 6 was drafted by Ling Jianming and Liao Chaohua. Section 2, Section 3, Section 4 and Section 9 of Chap. 7 were drafted by Yao Hailin. Section 7 and Section 16 of Chap. 7 were drafted by Zhang Liujun. Section 8 and Section 19 of Chap. 7 were drafted by Wu Lijian. Section 10 of Chap. 7 was drafted by Ding Xiaojun. Section 11 of Chap. 7 was drafted by Han Zhiqiang. Section 12 of Chap. 7 was drafted by Zhang Jinzhao. Section 13 and Section 15 of Chap. 7 were drafted by Chen Xiaoguang. Section 14 of Chap. 7 was drafted by Liu Jian.

Inquiries, suggestions, and comments are invited for further and continuous improvement in highway engineering, which should be sent to the daily management team in charge; viz. Wu Wanping (Add: No. 18, Chuangye Road, Economic, and Technological Development Zone, Wuhan, CCCC Second Highway Consultants Co., Ltd., Postal Code: 430056, Tel: 027-84214041, Fax: 027-84214068, E-mail: wanphx@263.net).

Editing Organization

CCCC Second Highway Consultants Co., Ltd.

Associate Editing Organizations

CCCC First Highway Consultants Co., Ltd.
China Merchants Chongqing Communications Technology Research & Design Institute Co., Ltd.
Tongji University
Chang'an University
China Academy of Transportation Science
Research Institute of Highway Ministry of Transport
Xinjiang Academy of Transportation Sciences
Institute of Rock and Soil Mechanics, Chinese Academy of Sciences

Contributors

Ma Lei
Chen Zhongping
Feng Shouzhong
Yang Jing
Yuan Guangyu
Fu Wei

Introduction to English Version

Standards reflect the achievement of civilization, provide common languages for technical communications and improve global connectivity. In recent years, Chinese government has been proactively implementing the standardization to stimulate innovation, coordination, greening and opening up for shared development in China and worldwide. In light of mutual development along the Silk Road Economic Belt and the 21st-Century Maritime Silk Road (so called the Belt and Road Initiative), the Ministry of Transport of the People's Republic of China organized translation and published international version of Chinese transportation industry standards and specifications to cope with the increasing demands for international cooperation in world transportation, achieve interconnected development and promote knowledge dispersion and experience sharing.

JTG is the designation referring to the standards and specifications of highway transportation industry, issued by Ministry of Transport of the People's Republic of China. It covers the standards and specifications in terms of technology, administration and service for whole process from highway planning through to highway maintenance. The criteria for safety, environment and economy assessment are also included.

The Chinese government has always attached critical importance to the construction of transportation infrastructure and makes great efforts to improve standards related to highway infrastructure design. In the 1980s, based on the original *Technical Standard of Highway Engineering* (JTJ 01—81), China began to formulate professional technical specifications for highway routes, substructures, pavements, bridges, and culverts. In February 1986, China announced and issued the first version of *Specifications for Design of Highway Subgrades* (JTJ 013—86). Thereafter, it released the first revision in 1995 (JTJ 013—95), the second revision in 2004 (JTG D30—2004) and the third revision in 2015 (JTG D30—2015). After nearly 40 years of technological development, a relatively complete technical specification system for highway substructure design has been established.

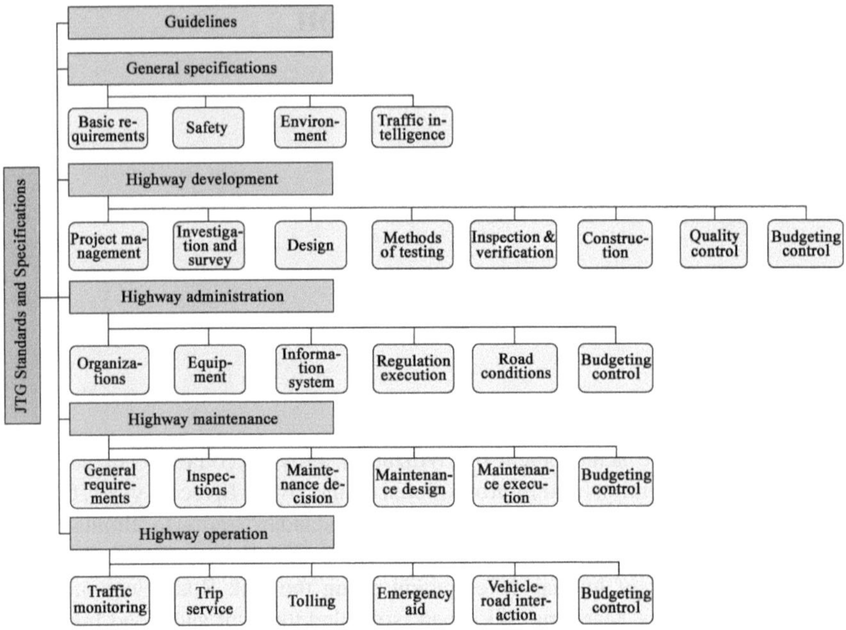

As is well acknowledged, China has a vast territory. In this case, China owned a complex and changeable climate, a sensitive and fragile environment and a complex topography and geology. It is widely distributed in soft soil, expansive soil, loess, saline soil, desert, permafrost, karst and other special rocks, as well as landslides, avalanches, mudslides, etc. Poor geology, high-filled embankment or deep-dug cut exists on roads in mountain area. In the process of highway construction, Chinese engineers and technicians resolved the critical technical issues of subgrade construction under various terrain and geological complex conditions, accumulated rich engineering experience, improved subgrade design theories and methods, as well as special subgrade construction and disasters prevention technology, etc. These experiences and achievements have been fully reflected in *Specifications for Design of Highway Subgrades* (JTG D30—2015). Therefore, the publication of this English edition is targeted at sharing Chinese engineering experience and technology with international counterparts and providing reference to the similar case.

Assuredly, with special attention, there exist slight differences in the highway construction division at China homeland and abroad. Therefore, the term 'subgrade' in foreign countries corresponds to the term 'subgrade structure' in China. It refers to the filling soil and its structure under the pavement layers. Its function is to bear the traffic load from the road surface and various natural factors. The 'substructure' in the Specifications includes the whole parts of highway trapezoidal cross section. That is to say, it is generally referred to as general facilities; not only the subgrade, but also the slope protection, drainage and retaining wall are included. In order to accurately express the concept of generalized highway cross section in the Specifications, the word 'substructure' was chosen in this English version.

The revision and updating of the English version were conducted by China Communications Press Co., Ltd., under the authorization of Ministry of Transport, China, and approved by Highway Department of MoT China.

Gratitude is given here to Mr. Wu Wanping, the editor in charge of the Chinese version for the valuable assistance and suggestions during the editing and approving of the English version.

Comments, suggestions and inquiries are welcome and should be addressed to editing organization in charge for English version (Address: No. 3, Waiguanxiejie, Andingmenwai, Chaoyang District, Beijing, P.R. China; Postal Code: 100011; E-mail: wym@ccpress.com.cn). The feedbacks shall be taken into account in future editions.

Editing organization for English version: China Communications Press Co., Ltd.

Associate editing organization for English version: Southeast University

Chief editors for English version: Huang Xiaoming, Han Min

Associate editors for English version: Wu Youming, Yang Jun, Ma Tao, Yu Bin, Zhang Junhui, Wang Fei, Ding Yao

Chief reviewer for English version: Yao Haidong

Associate reviewers for English version: Zhang Huiyu, Shen Shihui, Dr. Wen-Huei Phillip Yen

Public Notice No. 66; The People's Republic of China; Ministry of Transport

Public Notice on Issuing the English Version of Specifications for Design of Highway Substructure

The English version of JTG D30—2015 *Specifications for Design of Highway Substructure* is issued hereby for international cooperation and standardization of the highway transportation industry (this set of specifications hereinafter referred to as 'the *Specifications*').

The general administration and final interpretation of the English version of the *Specifications* belong to Ministry of Transport, while particular interpretation for application and routine administration of the English version of the *Specifications* shall be provided by China Communications Press Co., Ltd.

In event of any ambiguity or discrepancies between the English version and the Chinese version of the *Specifications*, the Chinese version should be referred to and accepted.

Comments, suggestions and inquiries are welcome and should be addressed to China Communications Press Co., Ltd. (Address: No. 3, Waiguanxiejie, Andingmenwai, Chaoyang District, Beijing; Postal Code: 100011; E-mail: wym@ccpress.com.cn). The feedback will be considered in future editions.

It is hereby announced.

Ministry of Transport of the People's Republic of China
August 14, 2020

Public Notice No. 11; The People's Republic of China; Ministry of Transport

Public Notice on Issuing the Specifications for Design of Highway Substructure

Hereby, the *Specifications for Design of Highway Substructure* (JTG D30—2015) is issued as one of Highway Engineering Industry Standards, to become effective on May 1, 2015. The former edition of the *Specifications for Design of Highway* Substructure (JTG D30—2004), and its English and French versions, shall be superseded from the same date.

The general administration and final interpretation of the *Specifications for Design of Highway Substructure* (JTG D30—2015) belong to Ministry of Transport, while particular interpretation for application and routine administration of the *Specifications* shall be provided by CCCC Second Highway Consultants Co., Ltd.

Comments, suggestions and inquiries are welcome and shall be addressed to CCCC Second Highway Consultants Co., Ltd. (No. 18, Chuangye Avenue, Wuhan Economy and Technique Development District, Hubei Province, Postal Code: 430056). The feedbacks will be considered in future editions.

It is hereby announced.

Ministry of Transport of the People's Republic of China
February 15, 2015

Contents

Chapter 1
General Provisions

1.1 The *Specifications* has been formulated to establish uniformity for technical standards of highway substructure design and to make substructure engineering design meet the requirements of safety, reliability, technical advancements, economy and rationality.

1.2 The *Specifications* is applicable to the substructure design of the new construction and expansion of all classified highways.

1.3 Substructure shall have a sufficient strength, stability and durability.

1.4 Geological investigation and geotechnical testing along the highway route shall be performed to ascertain the hydrogeological and geological conditions and obtain physical–mechanical parameters of rocks and soils as required for a substructure design.

1.5 A substructure design shall be conducted integratedly in accordance with the highway function and class, under the guidance of 'fit to local conditions, native material sources, minimized land using and environmental conservation', and then throughout evaluation on technical and economical options, to work out an acceptable solution for further comprehensive development.

1.6 The substructure design shall carry out the national policies in relation to technology and economy. New techniques, modern structures, innovative materials and advanced technologies shall be encouraged and adopted proactively and prudently.

1.7 Not only shall the requirements stipulated in the *Specifications* be complied with in substructure design, but also those in the current relevant national and industry standards and specifications shall be followed as well.

W. Wu, *Specifications for Design of Highway Substructure*, JTG Standards and Specifications, https://doi.org/10.1007/978-981-96-2313-6_1

Chapter 2
Terms and Symbols

2.1 Terms

2.1.1 Roadway substructure

A band-shaped structure built at location of highway route in compliance with certain technical requirements, that performs as the base of pavements and supports the traffic loads transmitted by pavements.

2.1.2 Roadbed

A part of substructure, 0.80 m or 1.20 m beneath the bottom of the pavement, that can be divided into two layers, an upper roadbed and a lower roadbed. A upper roadbed is 0.3 m in thickness, while a lower roadbed is either 0.5 m in thickness for the highways under the light, medium or heavy traffic or 0.9 m for the highways under the extra and extremely heavy traffic.

2.1.3 Embankment

The part of filled substructure beneath the bottom of roadbed and above original ground, that can be divided into two layers, upper embankment and lower embankment. Upper embankment is beneath the bottom of roadbed down up to 0.7 m in thickness. Lower embankment refers to the substructure fill below the upper embankment.

2.1.4 Cutting

An excavated substructure below the original ground level.

2.1.5 Substructure work zone

The region under roadbed which the stress ratio produced by traffic load with the pavement substructure and subgrade gravity is larger than 0.1.

2.1.6 Low embankment

W. Wu, *Specifications for Design of Highway Substructure*, JTG Standards and Specifications, https://doi.org/10.1007/978-981-96-2313-6_2

An embankment with a height of fill less than the depth of the substructure work zone.

2.1.7 High embankment

An embankment with a slope height larger than 20 m.

2.1.8 Steep slope embankment

An embankment with a ground slope steeper than 1:2.5.

2.1.9 Deep cutting

A soil cutting with a height larger than 20 m or a rock cutting with a height larger than 30 m.

2.1.10 Rockfill embankment

An embankment filled with materials containing more than 70% stone with a diameter greater than 40 mm.

2.1.11 Degree of compaction

The ratio of the dry density of compacted road materials to their standard maximum dry density, which is expressed as a percentage.

2.1.12 Special substructure

A substructure that needs to be designed specifically, which is always located in sections with special soils and rocks and harmful geological conditions and strongly influenced by water, climate and other natural factors.

2.1.13 Soft soil

A kind of fine grained soils with a high rate of natural water content, large void ratio, high compressibility, low shear strength, which refers to soft and weak soils including soft clay, mucky soil, sludge, cumulosol and peat in general.

2.1.14 Collapsibility loess

A loess that is rapidly destroyed and significantly sunk under a self-weight stress or certain pressure after being soaked.

2.1.15 Laterite

A brown red silt or clay formed by carbonate rocks weathering in humid climates.

2.1.16 High liquid limit soil

A kind of fine grained soils with a liquid limit larger than 50% (tested by 100 g cone).

2.1.17 Expansive soil

A kind of clays of a high plasticity containing hydrophilic minerals, which swells when watered and shrinks when dried out obviously.

2.1.18 Saline soil

A soil with a soluble salt content greater than the specified value.

2.1.19 Permafrost

A soil or rock frozen for two or over two years, which is in temperature below 0 °C and contains ice.

2.1.20 Seasonally frozen soil

A soil freezes and melts with seasons.

2.1.21 Landslide

A geological phenomenon of rocks or soils on slopes sliding along a slip belt or slip surface under the influence of natural or man-made factors.

2.1.22 Rock fall

A geological phenomenon of rocks or soils on high and steep slopes collapsing, dumping or falling by gravity.

2.1.23 Debris flow

An intermittent flood with large amounts of muds, sands and rocks.

2.1.24 Karst

All kinds of geological phenomena and forms, which results from the corrosion of soluble rocks by water for a long time.

2.1.25 Mined-out area

The area where the underground solid mineral deposit is extracted and the area or scope where the surface deformation and failure appear, which are caused by the displacement, cracking, broken down till the overall sinking and bending due to the surrounding rock losing stability.

2.1.26 Retaining wall

A structure to withstand a lateral earth pressure.

2.1.27 Slide-resistance pile

A pile under a lateral force to resist a soil pressure or slide down force.

2.1.28 Soil nailing

A supporting structure which maintains the stability of slopes by setting reinforcing screws in soil slopes or broken soft rock slopes.

2.1.29 Prestressed anchor

Consists of an anchor head, tendon and anchor body, which is a supporting structure which reinforces soils and rocks mass by applying a prestressing force to tendons.

2.1.30 Flexible supporting structure

A kind of supporting forms of substructure, which limits the large deformation of substructure slopes and permits a certain deformation of the structure.

2.2 Symbols

c—Cohesive strength of soil and rock;
E_0—Resilient modulus of substructure;
E_s—Compressed modulus of soil;
F_s—Stability coefficient of substructure;
K—Safety factor;
K_c—Anti-sliding stability coefficient of retaining wall;
K_0—Anti-overturning stability coefficient of retaining wall;
φ—Internal friction angle of substructure filling, ground soil, soil and rock of slope;
γ—Unit weight of substructure filling, ground soil.

Chapter 3
General Substructure

3.1 General Requirements

3.1.1 Before design, comprehensive design data about climate, hydrology, topography, geomorphology, seismicity, construction materials and so forth shall be collected along the route, to conduct the investigation and test works about geology and substructure filling materials along the route, and to clarify the geotechnical properties, thicknesses, space distribution characteristics of the soils or rocks and their relevant mechanical parameters.

3.1.2 Substructure design should avoid high fill and deep cut. When they are unavoidable, substructure design comparisons should be made based jointly on the route scheme, structures like bridges and tunnels or to adopt separated substructures or not if the fill height (in center) of the substructure exceeds 20 m or the cut depth (in center) of the substructure exceeds 30 m.

3.1.3 Elevation of substructure along the river side or could-be soaked substructure area shall be greater than the summation of the calculated height in accordance with the regulations of design flood frequency in Table 3.1, the backwater height, the wave height, and an extra height of 0.5 m for safety.

3.1.4 Substructure design shall select appropriate cross section type and slope gradient in accordance with local natural and engineering geologic conditions. Substructure along the riverside should not invade river course. Necessary protective facilities shall be used in accordance with erosion conditions. Waste soil of substructure shall be properly handled in order to avoid riverbed blocking, river diversion, and also to prevent destroying structures, farmland and housing along the river.

3.1.5 Fill material for substructure shall satisfy the requirements of strength and resilient modulus. Earthwork allocation design shall make technical and economic comparisons among plans such as excavated material used for filling, concentrating borrow or spoil, and modifying fill materiality, to thoroughly utilize the excavated materials and save land.

W. Wu, *Specifications for Design of Highway Substructure*, JTG Standards and Specifications, https://doi.org/10.1007/978-981-96-2313-6_3

Table 3.1 Design flood frequency for highway design

Highway class	Motorway	Class-1 highway	Class-2 highway	Class-3 highway	Class-4 highway
Design flood frequency	1/100	1/100	1/50	1/25	Determined according to site circumstances

Note For such a highway as the only pass way within a region, the design flood frequency of its embankment may be elevated by one level

3.1.6 Substructure design shall control the settlement after construction. Comprehensive measures shall be adopted in weak substructure, junctures between substructure and bridge or culvert structures, cut to fill location, high-fill embankment, steep embankment and so forth, in order to prevent the uneven deformation.

3.1.7 Substructure design shall take consideration of the influence from water and frost. It is necessary to set efficient drainage systems or anti-freeze facilities and other necessary substructure protection works.

3.1.8 Dynamic design shall be used for high-fill embankment, steep embankment and deep cutting and others of Motorway and Class-1 highway. Dynamic design must be based on the complete construction drawings and is applicable for substructure construction phase.

3.2 Roadbed

3.2.1 The thickness of roadbed shall be determined by traffic volume and the axle-louds compositions. Special calculation shall be carried out to determine substructure's working zone depth and the roadbed thickness for highway with special axle-loads.

3.2.2 Filling material for roadbed shall be uniform. The minimum California Bearing Ratio (CBR) shall meet the requirements in Table 3.2.

3.2.3 Roadbed shall be compacted layer by layer, and in accordance with:

1 The maximum size of filling material shall be less than 100 mm.

2 The degree of compaction shall be in accordance with Table 3.3.

3 The transverse slope on the top of roadbed shall be consistent with the crown slope.

3.2.4 Substructure design shall use substructure resilient modulus as design index and vertical compressive strain measured on the top of roadbed as checking index, and meanwhile be in accordance with:

1 When the substructure is at the equilibrium humidity state, the roadbed resilient modulus shall not be less than the specified values of *Specifications for Design of Highway Asphalt Pavement* (JTG D50) and *Specifications for Design of Highway Cement Concrete Pavement* (JTG D40).

Table 3.2 The minimum California Bearing Ratio (CBR) requirement for roadbed materials

Embankment classification		Depth below the bottom of pavement (m)	Minimum CBR of fill material(%)		
			Motorway, Class-1 highway	Class-2 highway	Class-3 and 4 highways
Upper roadbed		0–0.3	8	6	5
Lower roadbed	Light, medium and heavy traffic	0.3–0.8	5	4	3
	Extra heavy and extreme heavy traffic	0.3–1.2	5	4	–

Note 1. The conditions for carrying CBR tests in the table shall meet the requirements in the *Test Methods of Soils for Highway Engineering* (JTG E40)

2. If demonstrated by testing, equilibrium moisture content may be used as a condition in CBR test for regions with annual precipitation less than 400 mm, or non-immersed embankment with efficient drainage. And by jointly taking local climate and traffic-load-class into consideration, CBR control standard for filling material shall be determined

Table 3.3 Degree of compaction requirement for roadbed

Project classification		Depth below the bottom of pavement (m)	Degree of compaction(%)		
			Motorway, Class-1 highway	Class-2 highway	Class-3 and 4 highways
Upper roadbed		0–0.3	≥96	≥95	≥94
Lower roadbed	Light, medium and heavy traffic	0.3–0.8	≥96	≥95	≥94
	Extra heavy and extreme heavy traffic	0.3–1.2	≥96	≥95	–

Note 1.The degree of compaction in the table is the maximum dry density determined by the heavy compaction test according to the *Test Methods of Soils for Highway Engineering* (JTG E40)

2. If asphalt or concrete pavement to be paved on Class-3 and 4 highways, the degree of compaction shall choose the values for Class-2 highway

2 The calculated value of the vertical compressive strain on the top of roadbed shall meet the permanent deformation control requirement for asphalt pavement.

3 Control of the vertical compressive strain on the top of roadbed for cement concrete pavement may not be required.

3.2.5 The resilient modules of substructure E_0 for newly constructed highway shall be determined by Eq. (3.2.5.1), and meet the requirement by Eq. (3.2.5.2):

$$E_0 = K_s K_\eta M_R \tag{3.2.5.1}$$

$$E_0 \geq [E_0] \tag{3.2.5.2}$$

where:

E_0—design value of substructure resilient modulus at the equilibrium humidity state (MPa);

$[E_0]$—required value of substructure resilient modulus for designing pavement structure (MPa), which shall be ascertained by the requirements of Clause 3.2.4;

M_R—dynamic resilient modulus of substructure at the standard state (MPa), which is acquired from Clause 3.2.6;

K_s—moisture adjust factor of substructure resilient modulus, defined as the ratio of resilient modulus at the equilibrium humidity state (moisture content) and at the standard state, which is acquired from Clause 3.2.7;

K_η—reduction factor of substructure soil modulus for conditions with the dry–wet cycle or freeze–thaw cycle, which is determined by the test. In the preliminary design, it may be determined according to the soil type and moisture loss for non-frost zone, and according to the freeze temperature and moisture content for seasonal frost zone. It takes a value from 0.7 to 0.95. For non-frost zone, it takes a lower value for silt soil and clay soil, as well as when the rate of moisture loss exceeds 30%; oppositely it takes the relative higher value; coarse-grained soil takes high numerical value. For seasonal frost zone, it takes the low value for silt soil and clay soil with freeze temperature below -15 °C or with high moisture content before freeze; oppositely, it takes the relative high numerical value; coarse-grained soil takes the high numerical value.

3.2.6 Resilient modulus of substructure at the standard state shall be determined by:

1 Resilient modules of substructure fill material shall be obtained via test in accordance with Appendix A.

2 When the test is not feasible, the reference resilient modulus may be taken from Table B.1 and B.2 in light of the soil classification and aggregate type in Appendix B.

3 In the preliminary design, according to Eqs. (3.2.6.1) and (3.2.6.2) the resilient modulus of fill material at the standard state may be estimated by the CBR value of fill material:

$$MR = 17.6CBR^{0.64} \quad (2 < CBR \leq 12) \tag{3.2.6.1}$$

$$M_R = 22.1CBR^{0.55} \quad (12 < CBR < 80) \tag{3.2.6.2}$$

3.2.7 The roadbed of newly constructed highway shall be in the dry or medium wet conditions. For substructure design, the humidity condition may be estimated and then the moisture adjust factor of resilient modulus may be determined by following methods:

1 The substructure type defined by its moisture content may be determined from Appendix C in light of the relative height of substructure, substructure soil classification and rising height of capillary water, and then equilibrium humidity of substructure may be estimated.

2 The moisture adjust factor of resilient modules of embankment may be determined in Appendix D.

3.2.8 When the humidity state of substructure, CBR of substructure fill material, resilient modulus and vertical compressive strain of roadbed cannot meet the requirements, improvement measures shall be taken in light of the climate, soil, ground water storage, resources and so forth. After performing the technical and economic comparisons, the following measures can be taken:

1 Replacement by coarse-grained soil or soil stabilized with inorganic binding materials may be performed with appropriate replacement depth.

2 Mixing fine-grained soil with sand, gravel, macadam, or stabilized with inorganic binding materials may be adopted. Treatment of fine-grained soil shall be designed with physical mechanical test to determine the binding material and content, as well as the performance indicators of substructure after treatment.

3 Measures including set of drainage cushion layer, capillary water separation layer, and underground drainage sewer etc. shall be taken for cutting substructure with poor hydrogeological condition or filling embankment at wet condition.

4 Sections with medium wet or wet conditions in seasonal frost zone for every highway class shall be conducted with anti-freeze check of the substructure structure. The check of substructure structure shall consider pavement structure. Anti-freeze cushion layer or insulation layer shall be used if necessary.

3.3 Fill Embankment

3.3.1 Elevation of embankment shall be in accordance with:

1 Meet the requirements of design flood frequency of substructure and the corresponding flood elevation for respective highway class.

2 Elevation of embankment should not be less than the critical elevation corresponding to medium wet condition of substructure.

3 Elevation of embankment should not be less than local embankment frozen depth for seasonal frost zone.

3.3.2 Elevation of embankment should be calculated by Eq. (3.3.2).

$$H_{op} = \mathrm{MAX}\{\ (h_{sw} - h_0) + h_w + h_{bw} + \Delta h, h_l + h_p, h_{wd} + h_p, h_f + h_p\} \tag{3.3.2}$$

where:

H_{op}—reasonable elevation of embankment (m);
h_{sw}—design flood elevation (m);
h_0—ground elevation (m);
h_w—wave attack height (m);

h_{bw}—backwater height (m);
Δh—safety height (m);
h_l—critical elevation of medium wet substructure (m);
h_p—pavement thickness (m);
h_{wd}—depth of substructure workaround (m);
h_f—substructure frozen depth of seasonal frost zone (m).

3.3.3 Embankment fill material shall be in accordance with:

1 Coarse grained soil such as well-graded gravelly soil and sandy soil should be preferentially used as fill material, and fill materials shall be with a maximum particle size smaller than 150 mm.

2 Peat, sludge, frozen soil, expansive soil, soil with organic matter and soil with excessive soluble salt content shall not be directly used for filling. Silty soil shall not be directly used as fill material for roadbed in the seasonal frost area and soaked embankment area.

The minimum strength of the material shall be in accordance with those specified in Table 3.4.

4 Fine-grain soil, with a liquid limit above 50% and plasticity index higher than 26, shall not be directly used as embankment fill material.

5 Immersed embankment, bridge abutment, culvert backfill, and backfill of retaining wall should be constructed using materials with good water permeability. When find-grained soil is used as fill in regions lack of permeable materials, inorganic binding materials may be used to stabilize.

Table 3.4 The minimum strength requirements of embankment filling material

Project classification		Depth below the bottom of pavement (m)	Minimum strength of filling material (CBR)(%)		
			Motorway, Class-1 highway	Class-2 highway	Class-3 and 4 highways
Upper embankment	Light, medium and heavy traffic	0.8–1.5	4	3	3
	Extra heavy and extreme heavy traffic	1.2–1.9	4	3	–
Lower embankment	Light, medium and heavy traffic	Below 1.5	3	2	2
	Extra heavy and extreme heavy traffic	Below 1.9			

Note 1. When the CBR value of embankment material cannot meet the listed requirements, measures of mixing lime or other stable materials may be taken

2. For the pavement of asphalt or concrete on Class-3 and 4 highways, the strength shall meet the specified value for Class-2 highway

3.3.4 Each layer of embankment material shall be compacted to a uniform and dense state. The degree of compaction shall be in accordance with those specified in Table 3.5.

3.3.5 The embankment slope form and slope rate shall be designed in accordance with consideration of the physical and mechanical properties of fill, slope height and geological conditions.

1 With good geological conditions and a slope height of no more than 20 m, the slope rate should not be more than the values specified in Table 3.6.

2 For an embankment with slope height more than 20 m, a stepped-type slope should be used but the slope shall be determined by the stability analysis based on the provision in Sect. 3.6, and an individual design shall be performed.

Table 3.5 Degree of compaction for embankment

Project classification		Depth below the bottom of pavement (m)	Degree of compaction (%)		
			Motorway, Class-1 highway	Class-2 highway	Class-3 and 4 highways
Upper embankment	Light, medium and heavy traffic	0.8–1.5	≥94	≥94	≥93
	Extra heavy and extreme heavy traffic	1.2–1.9	≥94	≥94	–
Lower embankment	Light, medium and heavy traffic	Below 1.5	≥93	≥92	≥90
	Extra heavy and extreme heavy traffic	Below 1.9			

Note 1. The degree of compaction in the table is the maximum dry density tested by the heavy compaction test according to the *Test Methods of Soils for Highway Engineering* (JTG E40)

2. For a pavement of asphalt or concrete on Class-3 and 4 highways, the degree of compaction shall meet the specified value for Class-2 highway
3. For an embankment that uses special fill such as fly ash and industrial waste material, or is in a special climate area, the degree of compaction may be appropriately reduced based on the test road demonstration with the premise of ensuring roadbed strength requirements

Table 3.6 Slope rate of embankment

Material classification	Slope	
	Top height ($H \leq 8$ m)	Lower height ($H \leq 12$ m)
Fine-grained soil	1:1.5	1:1.75
Coarse-grained soil	1:1.5	1:1.75
Over coarse-grained soil	1:1.3	1:1.5

3 The slope of soaked embankment below the design water level should not be steeper than 1:1.75.

3.3.6 The ground surface treatment on stable slope shall meet the following requirements:

1 When the slope of the regional ground is less than 1:5, roadbed may be directly filled on the natural ground after the removal of surface turf and humus soil. Stepped excavation shall be done when the existing ground has a cross slope within 1:5 to 1:2.5. Bench widths shall not be less than 2 m. The thinner topsoil on the bed rock should be removed before excavating steps while the thick and stable covering layers may remain.

2 When the ground cross slope is steeper than 1:2.5 on the steep slope section, it is required to check and calculate the stability of embankment as a whole to slide along embankment bottom and along the soft-weak layer under it. The sliding stability coefficient shall not be less than what is specified in Table 3.11. Otherwise, measures to improve the base condition or anti-slip measures such as retaining structure shall be applied.

3 When embankment stability affected by underground water, measures such as intercepting and discharging underground water or filling permeable materials at the bottom of the embankment shall be taken.

4 Ground surface shall be densely compacted. For general soil sections, the substrate compaction tested by the heavy duty method shall not be less than 90% for Motorway and Class-1 and 2 highways, and shall not be less than 85% for Class-3 and 4 highways. When the roadbed filling height is less than the total thickness of pavement and substructure, the ground surface soil shall be excavated and each backfill layer shall be densely compacted. The depth of the treatment shall not be less than the workspace depth of roadbed subjected to heavy vehicle loading.

5 For sections in rice fields, lakes, ponds and so forth, measures such as drainage, dredging, drying, replacement, reinforcement and mixing of inorganic materials shall be taken according to the actual conditions. For sections on soft ground, the treatment measures shall meet the requirements of Sect. 7.7.

3.3.7 For Motorway and Class-1 and 2 highways, the transition section shall be used at the junction of embankment and bridge abutment or lateral structures, and the roadbed compaction shall not be less than 96%. The length of transition section should be calculated by Eq. (3.3.7).

$$L = (2\text{–}3)H + (3\text{–}5) \tag{3.3.7}$$

where:

L—length of transition section (m);

H—height of embankment fill (m).

3.3.8 Cut-fill substructure on steep slope may adopt shoulder-protection, masonry or retaining wall in consideration of terrain and geology conditions. Structures such as bridge and suspension platform may be used in steep or unstable slope, where excavation should not be adopted. Half tunnel may be used in cliff area for Class-3 and 4 highways where the rock is of good integrity.

Table 3.7 Rate of masonry slope

Number	Masonry height (m)	Inner slope	Outer slope
1	≤5	1:0.3	1:0.5
2	≤10	1:0.5	1:0.67
3	≤15	1:0.6	1:0.75

3.3.9 The height of the shoulder-protection should not exceed 2 m. The top width shall not occupy the hard shoulder, carriage way or marginal strip.

3.3.10 Masonry substructure may be used in Class-3 and 4 highways, and in accordance with:

1 Masonry substructure shall be layed by anti-weathering local rubble and block stone, and its inner side shall be filled with stone.

2 Masonry substructure should not be used in any highly weathered rock or soft rock sections.

3 The masonry shall have a top width of no less than 0.8 m, its height should be no more than 15 m, and its base surface shall be tilted inward. The inner and outer slope rates of the masonry should not be more than those specified in Table 3.7.

3.3.11 When there are terrain restrictions or the substructure has insufficient stability, a toe-protection substructure may be used. The height of the toe-protection should not be higher than 5 m and the soaking toe-protection substructure shall be protected or strengthened.

3.4 Cutting Substructure

3.4.1 Soil cutting design shall be in accordance with:

1 The form and slope rate of the soil cutting slope shall be based on geological and hydrogeological conditions, slope height, drainage measures, construction methods, as well as stability investigation and mechanical analysis for both natural slopes and artificial slope. When the slope height is not higher than 20 m, the slope rate should not be steeper than what is specified in Table 3.8.

Table 3.8 Rate of soil cutting slope

Soil type		Slope rate
Clay, silty clay soil, silt soil with plasticity index greater than 3		1:1
Medium sand, coarse sand, gravel sand of medium compressibility		1:1.5
Pebble soil, crushed gravel soil, round gravel soil, breccia soil	Good cohesiveness and compaction	1:0.75
	Medium dense	1:1

Note For special soils such as loess, red clay, high liquid limit soil and expansive soil, the soil cutting slope form and rate shall be determined by the requirements of Chap. 7

2 When the height of cutting slope is more than 20 m, the slope form and rate shall be ascertained by the requirements of Sect. 3.7.

3.4.2 Rock cutting design shall be in accordance with:

1 The form and slope rate of the rock cutting slope shall be based on geological and hydrogeological conditions, slope height, construction methods, as well as an investigation of natural slopes and artificial slope. Stability analysis methods may be used to check when necessary. For a slope with height less than 30 m, slopes without weak, outward heeling structural plane refer to Appendix E to determine the types of rock. The slope rate may be determined according to what is specified in Table 3.9.

2 For rock slopes with a weak, outward heeling structural plane that are subject to heavy loads on the edge of the slope top, as well as the heights exceeding the defined limits in Table 3.9, the slope rate shall be determined through the stability analysis and calculation specified in Sect. 3.7.

3 Smooth blasting and pre-splitting blasting technology should be adopted for hard rock excavation roadbed.

4 For weak and loose rock cutting with a slope height exceeding 20 m, the technologies of layered excavation and protection and pre-reinforcing slope toe should be used.

3.4.3 High slope in cutting may be excavated into a broken line type slope or stepped slope according to the different types of soil and rock and the stability requirements, the stage for heaping soil and broken rock at the foot of the cutting slope shall be set at the outside of the side ditch, and its width should not be less than 1.0 m. The slope platform shall be set at the middle of the stepped slope and its width should not be less than 2 m.

3.4.4 The surface drainage system shall be set at the top, surface, toe and central platform of slope. All the surface drainage structure sizes shall be determined according to those specified in Sect. 4.2.

Table 3.9 Rock cutting slope rate

Rock type	Weathering degree of rock	Slope rate	
		$H < 15$ m	15 m $\leq H \leq$ 30 m
Class-1	Unweathered and slightly weathered	1:0.1–1:0.3	1:0.1–1:0.3
	Weakly weathered	1:0.1–1:0.3	1:0.3–1:0.5
Class-2	Unweathered and slightly weathered	1:0.1–1:0.3	1:0.3–1:0.5
	Weakly weathered	1:0.3–1:0.5	1:0.5–1:0.75
Class-3	Unweathered and slightly weathered	1:0.3–1:0.5	–
	Weakly weathered	1:0.5–1:0.75	–
Class-4	Weakly weathered	1:0.5–1:1	–
	Strongly weathered	1:0.75–1:1	–

Note 1. With reliable information and experience, it may not follow the requirements specified in this table

2. Strongly weathered rock of Class-4 includes weathered soft rock in different degrees

3.4.5 When the side slope has ponding wetland, groundwater seepage or emitting, the following measures shall be taken based upon the actual situation: setting up an underground seepage ditch, slope seepage ditch, up-dip drainage holes, or constructing drainage tunnels and other drainage facilities in the upstream along the perpendicular direction of the groundwater flow in order to intercept the underground water.

3.4.6 Slope protection types shall be determined according to the slope stability and surrounding environmental situation. Slope protection shall combine structure protection with flora protection. For an unstable slope, comprehensive supporting and retaining structures shall be adopted.

3.5 Boundary Region Treatment Between the Excavation and Fill Substructure

3.5.1 The construction of the cut-fill substructure filling area shall be in accordance with the relevant requirements of Sects. 3.3 and 3.6 of the *Specifications.* When necessary, impact-rolling or dynamic consolidation methods may be used to enhance the strength and density of the roadbed.

3.5.2 The design of the cut-fill roadbed filling area shall be in accordance with the relevant requirements in Sects. 3.4 and 3.7 of the *Specifications.*

3.5.3 When the excavation area is composed of soil or soft rock, the substructure soil that does not meet the requirements shall be over excavated, backfilled and compacted, or implemented with improvement treatment. Fill material with good permeability should be selected for filling area. When necessary, geogrid may be laid on the boundary region of the substructure at the fill and cut junction. When the excavation area is composed of hard rock, rock-fill roadbed should be used.

3.5.4 When the ground slope is steeper than 1:2.5, the substructure shall be designed in accordance with Sect. 3.6. For an unstable substructure, the supporting and retaining structures shall be set at the bottom of the embankment.

3.5.5 A perfect underground drainage system shall be set according to the exposure condition of the underground water and the soil properties. Excluding the longitudinal blind drain situated beneath the side ditch, lateral and longitudinal blind drains shall be set in the boundary region between fill and cut.

3.5.6 The transition section should be designed at the longitudinal boundary region between fill and cut.

3.6 High Embankment and Steep Slope Embankment

3.6.1 Individual investigation and design shall be performed if the embankment is of high and steep slope, of poor geology, or of special substructure.

3.6.2 Design of embankment with high slope or steep slope shall implement the principles of comprehensive design and consider specific situations design. The cross section of the embankment, drainage facilities, slope protection form, and the ground and embankment treatment shall be comprehensively designed with a full grasp of the hydrogeological conditions and the source and nature of the fill material at the site. When the actual situation changes, the design scheme shall be adjusted in time to ensure embankment stability.

3.6.3 For high slope and steep slope embankments, a geotechnical investigation shall be carried out to identify the type, position, thickness, distribution characteristics and physical–mechanical properties of the soil and to determine the depth and distribution of underground water, the bearing capacity of ground soil and to acquire physical and mechanical indexes in embankment design. The engineering geological investigation shall meet the requirements specified in the *Code for Highway Engineering Geological Investigation* (JTG C20).

3.6.4 Fill materials of high embankment and steep embankment shall meet the requirements of Clauses 3.2.2 and 3.3.3 of the *Specifications.* Embankment compaction shall meet the requirements of Clauses 3.2.3 and 3.3.4 of the *Specifications.*

3.6.5 The high embankment slope form and slope rate shall be determined according to stability calculations combined with economic and environmental factors based on terrain, engineering geological condition, slope height, and fill material properties etc. The stepped slope should be designed in cross sections of the high embankment.

3.6.6 A effective/comprehensive underground or surface drainage system and timely slope protection shall be set according to the exposure condition of the underground water and the soil properties.

3.6.7 Stability calculation shall be performed for high embankment and steep embankment for three scenarios:

1 Normal condition: the condition of high frequency or long duration after substructure opening to operation.

2 Abnormal condition I: under heavy rainstorm or continuous rainfall.

3 Abnormal condition II: under extreme loads such as earthquake.

3.6.8 The strength parameters for the high embankment stability analysis shall be determined based on fill material source, actual conditions on-site, laboratory tests of typical soil samples and consider the field situation. Test method shall be in accordance with:

1 The strength parameters c and φ values of embankment may be determined using quick direct shear test or a triaxial undrained shear test. The requirements for sample preparation and test methods of stability analysis in each stage are shown in Table 3.10. If the fill material is coarse grained soil or stone, a large triaxial testing apparatus shall be adopted.

2 The strength parameters C and Φ of the roadbed should be tested by a consolidated quick direct shear test or triaxial consolidated-undrained shear test.

3 When analyzing the sliding stability of the embankment along a sloping ground or weak layer, the soil sample from the controlling layer shall be chosen to acquire

Table 3.10 Embankment soil sample preparation requirement for strength parameter test

Working condition classification	Sample requirement	Applicable range
Normal	Use water content and density of embankment. When it is difficult to obtain them, or preliminary stability is analyzed, the density need to achieve the required values and the water content need to take the higher value corresponding to the required density on compaction curves	Newly constructed embankment
	Undisturbed soil from embankment	Existing embankment
Abnormal I	Ditto, but pre-saturated	Filled soil within the rain infiltration range
Abnormal II	Ditto	–

strength parameters c and φ by test in combination with the site conditions. The quick direct shear or triaxial unconsolidated-undrained shear test may be used. Saturated specimens shall be used for testing if there is underground water.

3.6.9 The stability of embankment body and the global stability of embankment and ground should be calculated and analyzed by the simplified Bishop method. The stability coefficient F_s is calculated according to the following Eq. (3.6.9.1).The calculation diagram is shown in Fig. 3.1. When the embankment is weak or soft, the calculation method of stability coefficient and the stability shall be in accordance with the requirements of Sect. 7.7.

$$F_s = \frac{\sum [c_i b_i + (W_i + Q_i)\tan\varphi_i]/m_{\alpha i}}{\sum (W_i + Q_i)\sin\alpha_i} \tag{3.6.9.1}$$

where:

F_s—stability coefficient of embankment;

b_i—horizontal width of slice i(m);

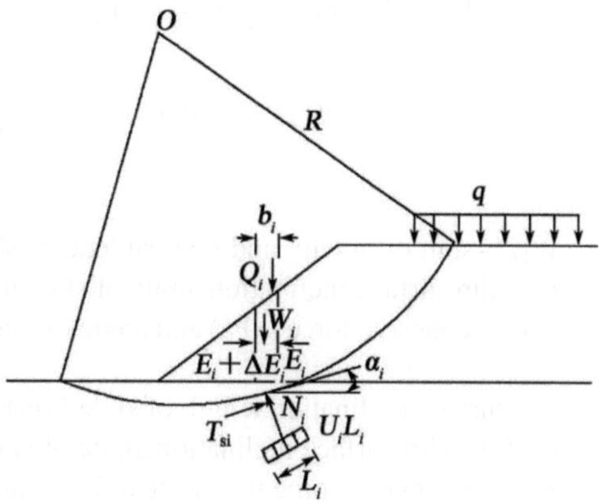

Fig. 3.1 The diagram depicts the simplified Bishop method

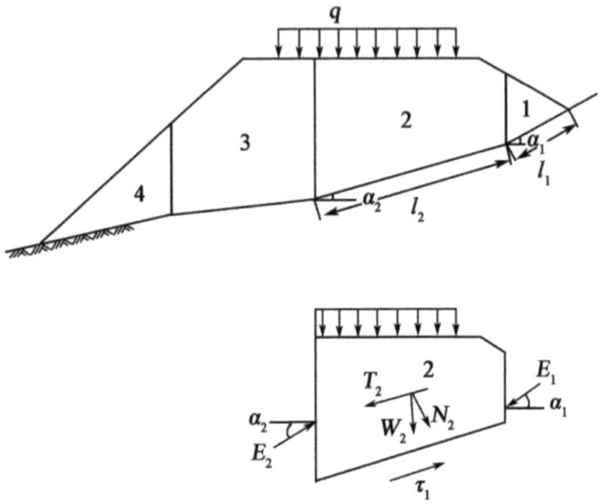

Fig. 3.2 The diagram depicts the unbalanced thrust method

α_i—slip surface inclination angle of slice i(°);

c_i, φ_i—cohesive force(kPa) and angle of internal friction(°) of the ground where the sliding arc of slice i is located;

$m_{\alpha i}$—coefficient, calculated by Eq. (3.6.9.2);

$$m_{\alpha i} = \cos\alpha_i + \frac{\sin\alpha_i \tan\varphi_i}{F_s} \tag{3.6.9.2}$$

W_i—gravity of slice i (kN);

Q_i—external force in vertical direction of slice i (kN).

3.6.10 The unbalanced thrust method may be used to analyze and calculate the sliding stability of embankment, which is along the sloping ground or at a soft and weak layer. Stability coefficient F_s may be calculated according to Eqs. (3.6.10.1) and (3.6.10.2). The calculation diagram is shown in Fig. 3.2.

$$E_i = W_{Qi}\sin\alpha_i - \frac{1}{F_s}(c_i l_i + W_{Qi}\cos\alpha_i \tan\varphi_i) + E_{i-1}\psi_{i-1} \tag{3.6.10.1}$$

$$\psi_{i-1} = \cos(\alpha_{i-1} - \alpha_i) - \frac{\tan\varphi_i}{F_s}\sin(\alpha_{i-1} - \alpha_i) \tag{3.6.10.2}$$

where:

W_{Qi}—sum of gravity and vertical load of slice I (kN);

α_i—slip surface inclination angle of slice i(°);

c_i,φ_i—cohesive force (kPa) and angle of internal friction(°) of the bottom of slice i;

l_i—surface inclination length of slice i (m);

α_{i-1}(°)—slip surface inclination angle of slice i–1 (°);

E_{i-1}—downslide force transmitted from slice i–1 to slice i (kN).

Use Eqs. (3.6.10.1) and (3.6.10.2) to determine the stability safety factor F_s by calculating slice one by one till the residual thrust of that slice is zero.

3.6.11 The stability safety factor, calculated according to the analysis of embankment stability, shall not be less than the values specified in Table 3.11. For Abnormal condition II, the analysis method of embankment stability and the stability safety factor shall be in accordance with the specifications of *Specification of Seismic Design for Highway Engineering* (JTG B02).

3.6.12 Foundation treatment shall meet the requirements of Clause 3.3.6. If the ground is at weak soil layers, the remediation design shall be done according to the requirements of Sect. 7.7. If the embankment stability safety factor is less than those specified in Table 3.11, control major measures to improve the base condition, use of retaining structures, or reinforcement measures shall be taken to assure embankment stability.

3.6.13 Settlement monitor of high and steep slope embankments shall be emphasized. When necessary, synthetical measures of impact-rolling or dynamic consolidation, laying geosynthetic material may be applied and should be reserved for one rainy season to reduce settlement after construction.

3.6.14 Construction monitor shall be conducted for high and steep slope embankments. Design shall define sections, the items for observation, the numbers, locations of observation points, requirements of observation and so forth. The observation items may be selected according to Appendix Table F-2. Observation duration shall be greater than 1 year after highway opening to traffic.

Table 3.11 Stability safety factor for high and steep embankment

Content of analysis	Strength index of embankment	Working condition classification	Stability safety factor	
			Class-2 and upper level	Class-3 and 4
Stability of embankment body, global stability of embankment and ground	The index is determined by the consolidated quick shear test or consolidated-undrained triaxial test	Normal	1.45	1.35
		Abnormal I	1.35	1.25
	The index is determined by the quick shear test	Normal	1.35	1.30
		Abnormal I	1.25	1.15
The sliding stability of embankment along the sloping ground or weak layer zone	–	Normal	1.30	1.25
		Abnormal I	1.20	1.15

Note For important section of Class-3 and 4 in region with only one highway, stability safety factor may be used with Class-2 standard

3.7 High Slope in Cutting

3.7.1 Individual investigation and design shall be performed for high excavation or excavation slope in poor technological conditions.

3.7.2 The slope engineering exploration should be performed by comprehensive means of drilling, pit (costean, trench) exploration and combined with geophysical explorations. When necessary, an adit exploratory may be supplemented. The slope engineering exploration shall meet the requirements specified in the *Code for Highway Engineering Geological Investigation* (JTG C20) and the following contents shall be identified:

1 Terrain and geomorphology characteristics.

2 The type, origin, physicochemical properties, weathering degree, structural integrity and layer thickness of the mass of soil or rock.

3 The physical and mechanical properties of saturated natural mass of soil or rock, (such as unit weight γ, strength parameters c and φ, etc.).

4 The main structural plane features (especially the weak structural plane), combination relation, mechanical properties and its relationship with free face.

5 The meteorological, hydrological and hydrogeological conditions.

6 The scope, properties and distribution of adverse geologic phenomena.

7 The load, structure, foundation type, buried depth and stability of the building near the top of slope.

8 Surface runoff form and its impact on slope.

3.7.3 Mechanical parameters of rock and soil slope may be determined:

1 The shear strength of rock mass should be determined by an in-situ soil test. The test shall comply with the requirements specified in the current national standard of *Standard for Tests Method of Engineering Rock Mass* (GB/T 50,266). When the conditions are not permitted for testing, the shear strength may be comprehensively determined using the *Standard for Engineering Classification of Rock Mass* (GB/T 50,218), Table 3.12 and inverse calculation analysis method.

2 Structural surface bonding degree of the rock mass may be determined according to Table 3.13.

3 Standard values of the slope rock mass performance index may be determined according to the local experience. The important slope shall be determined by a test.

4 The rock mass internal friction angle may be calculated by using standard values of rock internal friction angle multiplied by reduction factors listed in Table 3.14 according to the degree of development of rock fractures.

5 The physical parameters of soil should be fully ascertained by an in-situ shear test, laboratory test of undisturbed soil sample and analysis of reverse calculation method, etc.

6 When calculating soil slope using the principle of merging calculation of water and soil, the soil below groundwater level should adopt a shear strength index tested by the dead weight consolidated-undrained triaxial test. When using the principle of separate calculation of water and soil, the soil below groundwater level should adopt an effective shear strength index of soil.

Table 3.12 Standard values of structural shear strength index

Classification of structural surface		Degree of bonding at Binding degree of structural planes	Internal friction angle φ (°)	Cohesion c (MPa)
Hard structural surface	1	Good	> 35	> 0.13
	2	Average	35–27	0.13–0.09
	3	Poor	27–18	0.09–0.05
Weak structural surface	4	Very poor	18–12	0.05–0.02
	5	Extremely poor (such as mud layer)	Determined according to local experience	

Note 1. The time effect of structural surface has been taken into account for values in the table
2. Soft and extreme-soft rock takes the low values in the table
3. Structural surface of rock mass with poor connectivity takes high numerical values in the table
4. Soaking structural surface of rock mass takes the low values in the table

Table 3.13 Bonding degree of structural surface

Bonding degree	Structural surface features
Good	The fracture of rock mass is less than 1 mm, well cemented and unfilled; the fracture of rock mass is between 1–3 mm, and silica or ferruginous cemented
Average	The fracture of rock mass is between 1–3 mm, and calcite-cemented; the fracture of rock mass is greater than 3 mm, calcium-cemented, and surface is rough
Poor	The fracture of rock mass is between 1–3 mm, uncemented and surface is flat; the fracture of rock mass is greater than 3 mm, and filled by debris or mud plus debris
Extremely poor (such as argillized layer)	The surface is flat and smooth, cemented; mud or mud with debris filled, the thickness of the filling is greater than undulating difference; continuously distributed argillized seam; uncemented or strongly weathered small fault's fracture zones

Table 3.14 Reduction factors of internal friction angle of slope rock mass

Characteristic of slope rock mass	Reduction factor of internal friction angle	Characteristic of slope rock mass	Reduction factor of internal friction angle
Undeveloped cracks	0.90–0.95	Developed cracks	0.80–0.85
Moderate developed cracks	0.85–0.90	Cataclastic texture	0.75–0.80

3.7.4 Evaluation of slope stability shall follow the principle of 'qualitative analysis as the base and quantitative calculation as the tool'. Calculation of slope stability shall take the slope geological condition or emerged deformation damage sign into account to determine the probable failure mode and stability state.

3.7.5 Slope stability calculation shall take slope form and the probable failure modes into account, which is determined according to the following methods:

1 With a large scale cataclastic texture, the rock slope and soil slope should be calculated by using the simplified Bishop method.

2 A slope that possibly would be linearly broken should be calculated by using the plane sliding surface analytical method.

3 A slope that possibly would be broken on fold-line should be calculated by using the unbalanced thrust method.

4 The complex rock slope may be calculated by cooperating with the analysis by the stereographic projection method, proportionary stereographic projection method and wedge sliding surface method.

5 If the slope failure mechanism is complicated, it should be analyzed in combination with numerical analysis.

3.7.6 Slope stability analysis shall be divided into one of three working conditions. The impact of freeze–thaw shall be considered for seasonal frost soil slope.

1 Normal working condition: slope in a natural working condition;

2 Abnormal working condition I: slope with heavy rainstorm or continuous rainfall conditions;

3 Abnormal working condition II: slope under loads such as earthquake condition.

3.7.7 The safety factor of slope stability by calculation shall not less than the safety factors specified in Table 3.15. For Abnormal working condition II, the analysis method of cutting slope stability and the stability safety factor shall be in accordance with the specifications of *Specification of Seismic Design for Highway Engineering* (JTG B02).

3.7.8 Cutting slope should be excavated accordingly into a fold line type slope or stepped slope. A bench, with a width should not be less than 2.0 m, shall be set in the middle of the stepped slope. For a hard rock section, the slope may be built without the benches and the slope rate may be comprehensively determined according to slope stability analysis and investigation of artificial slopes and natural slopes of nearby existing engineering.

3.7.9 The design of slope protection shall take comprehensive measures of engineering protection and flora protection according to the geological and environmental conditions, height of the slope and class of highway. For a poor stability slope,

Table 3.15 Safety factors for cutting slope

Working condition	Safety factor for cutting slope	
	Motorway, Class-1 highway	Class-2 and lower highway
Normal working condition	1.20–1.30	1.15–1.25
Abnormal working condition I	1.10–1.20	1.05–1.15

Note 1. The safety factor for cutting slope with complicated geological condition or severe harm after damage takes high numerical value; Safety factor for cutting slope with simple or light harm after damage may take low value

2. When there are important structures (bridge, tunnel, high-voltage power transmission tower, and oil–gas pipeline etc.), village and school locating within the influencing range of cutting slope after damage, the safety factor can take high numerical value
3. Temporal safety factor for slope construction shall not be less than 1.05

comprehensive supporting and retaining structures shall be built and technology such as stratified excavation, stratified stabilization and slope toes pre-reinforcing shall be used.

3.7.10 An effective slope surface and underground drainage system shall be set to discharge surface water and underground water in time. The dimensions of drainage facilities shall be determined in accordance with Sects. 4.2 and 4.3. Insulation measure shall be applied to the outlet of underground drainage system when underground water is abundant in seasonal frost soil slope.

3.7.11 Construction monitor shall be carried out in high excavated slopes of Motorway and Class-1 highway and excavated slopes in adverse geological and special soil and rock sections. Design shall define sections, the items for observation, the numbers, locations of observation points, requirements of observation and so forth. The observation items may be selected according to Appendix Tables F-1 and F-3. Observation duration shall be greater than 1 year after highway opening to traffic.

3.8 Rock Embankment

3.8.1 Design of rock embankment shall be in accordance with:

1 Hard rock and medium hard rock may be as fill materials of roadbed. Soft rock may be used as fill materials of embankment and shall not be used as fill materials of roadbed. Expansive rock, soluble rock, collapsible rock and salinization rock shall not be used as fill materials of embankment.

2 The cross section, structure and drainage shall be effectively designed to ensure sufficient strength and stability of the rock embankment.

3 Before construction, the appropriate layer thickness, compaction techniques and quality control standards shall be determined by the test section.

3.8.2 According to the compressive strength index of saturated rock, fill rock may be classified as what is specified in Table 3.16.

3.8.3 Different filling thickness and compaction control standards shall be used for different strengths of stone. Porosity should be used as a control index for the rock embankment compaction quality standard; it should satisfy the requirements of Table 3.17, 3.18 and 3.19. Compaction may be jointly controlled by porosity percentage, compaction settlement difference or construction parameters.

3.8.4 The last spreading thickness of the rock-fill embankment shall not be greater than 400 mm. The aggregate size shall be smaller than 150 mm, among which the content of fines that are smaller than 5 mm shall not be less than 30%. The spreading layer shall be free of obvious pore and void. When the upper part of rock embankment is filled with other materials, geotextile may be set as the separation layer if necessary.

3.8.5 The type of rock-fill embankment cross section may be the same as the type of soil embankment cross section. The slope rate of rock-fill embankment should not be steeper than the specification in Table 3.20. The side wall may use rubble masonry and masonry's thickness should be 1–2 m. The size of rubble shall not be less than

Table 3.16 Rock classification

Rock classification	Uniaxial saturated comprehensive strength (MPa)	Representative rock
Hard rock	≥60	1. Granite, diorite, basalt and other igneous rocks; 2. Conglomerate with silica and iron cementation, sandstone, limestone, dolomite and other sedimentary rocks; 3. Gneiss, quartzite, marble, slate, schist and other metamorphic rocks
Medium hard rock	30–60	
Soft rock	5–30	1. Tuff and other volcanic rocks; 2. Muddy gravel, sand, shale, mudstone and other sedimentary rocks; 3. Mica schist, phyllite and other metamorphic rocks

Table 3.17 Compaction quality control standards for hard stone

Layer	Depth below the bottom of pavement (m)	Paving thickness (mm)	Maximum size (mm)	Compacted dry density (kg/m^3)	Percentage of porosity (%)
Upper embankment	0.80–1.50 (1.20–1.90)	≤400	Less than 2/3 of layer thickness	Determined by test	≤23
Lower embankment	>1.50 (>1.90)	≤600	Less than 2/3 of layer thickness	Determined by test	≤25

Note Numbers in parenthesis in the 'depth below the bottom of pavement' column represent depth ranges of upper embankment and lower embankment for extra and extreme heavy traffic

Table 3.18 Compaction quality control standards for mid-hard stone

Zoning	Depth following the bottom of pavement (m)	Paving thickness (mm)	Maximum size (mm)	Compacted dry density (kg/m^3)	Percentage of porosity (%)
Upper embankment	0.80–1.50 (1.20–1.90)	≤400	Less than 2/3 of layer thickness	Determined by test	≤ 22
Lower embankment	>1.50 (>1.90)	≤500	Less than 2/3 of layer thickness	Determined by test	≤ 24

Note Numbers in parenthesis in the 'depth below the bottom of pavement' column represent depth ranges of upper embankment and lower embankment for extra and extreme heavy traffic

Table 3.19 Compaction quality control standards for soft stone

Zoning	Depth below the bottom of pavement (m)	Paving thickness (mm)	Maximum size (mm)	Compacted dry density (kg/m^3)	Percentage of porosity (%)
Upper embankment	0.80–1.50 (1.20–1.90)	≤300	Less than layer thickness	Determined by test	≤ 20
Lower embankment	>1.50 (>1.90)	≤400	Less than layer thickness	Determined by test	≤ 22

Note Numbers in parenthesis in the 'depth below the bottom of pavement' column represent depth ranges of upper embankment and lower embankment for extra and extreme heavy traffic

Table 3.20 Slope rate of rock-fill embankment

Stone type	Slope height (m)			Slope rate	
	Overall height	Upper height	Lower height	Upper	Lower
Hard rock	20	8	12	1:1.1	1:1.3
Medium hard rock	20	8	12	1:1.3	1:1.5
Soft rock	20	8	12	1:1.5	1:1.75

300 mm. When the slope is high, a platform with a width of 1–3 m may be used in the middle part of slope.

3.8.6 If weathered rock and soft rock are used to build the rock-fill embankment, hard rock gravel or other qualified fill material shall be used in roadbed. Measures such as wrapping closure or reinforcement of embankment side, set drainage cushion layer at the bottom, and anti-infiltration layer at the top shall be adopted to prevent the moisture deformation of rock-fill embankment.

3.8.7 Rock-fill embankment structural design on soft ground shall be integrated with soft soil ground treatment.

3.9 Light Weight Material Embankment

3.9.1 Light weight material may be used in embankment to reduce gravity or soil pressure. It can be used in embankment on soft subgrade, backfill of retaining wall structures, embankment extension, repair of sinking or unstable embankment and so forth. It should not be used in flooded region.

3.9.2 Light weight material embankment design shall take design purpose, load level, terrain and geological conditions, environmental condition and geometric parameters of substructure into consideration. Technical and economic comparisons shall be carried out to select proper light material type, substructure structure and cross-section profile, and to determine design parameter of the material.

3.9.3 Light material as embankment fill shall be in accordance with:

1 Geotechnical plastic foam (such as EPS) should not have the density less than 20 kg/m^3, the compressive strength less than 110 kPa at 10% compressive strain, the bending strength less than 3.5 MPa, and the 7 d volumetric water absorption rate higher than 1.5%. Compressive strength of geotechnical plastic foam bulk beneath special positions such as bridge end transition slab shall not be less than 250 kPa. Flame-retarding geotechnical plastic foam bulk shall be used nearby buildings that have anti-fire requirement.

2 Light foam soil shall not have a construction wet unit weight less than 5.0 kN/m^3, and should not be greater than 11.0 kN/m^3. The flow number should be 170–190 mm and the unconfined compressive strength shall be in accordance with the specifications in Table 3.21. If required by project or constrained by environment, the anti-freeze index of light foam soil may be determined by the test.

3 Fly ash used in the embankment of motorway and class-1 highway should have an ignition loss less than 20%. Otherwise, comparison test shall be carried out to demonstrate the usability.

3.9.4 Design of light material embankment shall be in accordance with:

1 Effective protection measure shall be taken for light material embankment to avoid the direct exposure of light material. Substructure cross-section profile may adopt upright embankment with supporting structures or inclined embankment with wrapping. The fill depth of light material shall be determined by post-construction settlement.

2 A transition section of the stepped form shall be adopted between light material embankment and regular soil filled embankment. The height of the step should be 0.5–1.0 m and the slope ratio should be 1:1–1:2.

3 Embankment stability and settlement calculation shall be carried out for light material embankment design. Post-construction settlement of newly constructed embankment shall be in accordance with the requirement of Sect. 7.7; reorganization

Table 3.21 Unconfined compressive strength of light foam soil in substructure

Zone of substructure		Unconfined compressive strength (MPa)	
		Motorway, Class-1 highway	Class-2 highway and below
Road bed	Light, medium and heavy traffic	≥0.8	≥0.6
	Extra heavy and extreme heavy traffic	≥1.0	
Upper embankment, lower embankment		≥0.6	≥0.5
Replacement of subgrade soil		>0.4	

Note 1. The unconfined compressive strength refer to the cubic compressive strength of 28 d curing and side length of 100 mm

2. For roadbed of Motorway with extra and extreme heavy traffic and Class-1 highway, light foam soil should be mixed with sand. The flow number should be 150–170 mm and the mass ratio of sand and cement should be 0.5–2.0

and expansion of embankment shall be in accordance the requirement of Sect. 6.4; embankment stability shall be in accordance with the requirement of Sects. 3.6 and 7.7.

4 When the light material filled zone is lower than the underground water level or inundated, floating resistance stability shall be checked with Eq. (3.9.4). When coefficient of floating resistance stability is smaller than the safe value, measures such as adjust the thickness of light material filled zone or increase the weight of filled soil shall be adopted. Safety coefficient of light material embankment should be 1.1–1.5. Safety coefficient of light foam material embankment should be 1.05–1.15. When the probability of highest underwater level or flood level reaches the light material filled zone is low, safety coefficient takes low value.

$$F_{\mathrm{f}} = \frac{\Sigma \gamma_i h_i}{\gamma_{\mathrm{w}} h_{j\mathrm{w}}} \tag{3.9.4}$$

where:

F_{f}—coefficient of floating resistance stability;

γ_i—unit weight of each layer (kN/m^3);

h_i—thickness of each layer (m);

γ_{w}—unit weight of water (kN/m^3);

$h_{j\mathrm{w}}$—soaked depth of embankment (m).

3.9.5 Design of geotechnical plastic foam light material embankment shall be in accordance with:

1 Pouring reinforced concrete plate and impervious geotextile shall be adopted between geotechnical plastic foam and pavement, and between geotechnical plastic foam bulk of multi-layers at an interval of 2–3 m or 4–6 layers with consideration of load impacts from vehicle and substructure and pavement weight. The thickness of pouring reinforced concrete plate should be 0.10–0.15 m.

2 Sand-gravel cushion layer shall be used at the bottom of geotechnical plastic foam bulk, and its thickness should be 0.2–0.3 m. When necessary, pervious geotextile may be used at the up and low interface.

3 Compressive strength of material shall be checked for geotechnical plastic foam embankment design. The stress on the surface of geotechnical plastic foam bulk (refer to Fig. 3.3) caused by weight of pavement and reinforced concrete plate and moving load may be calculated with Eq. (3.9.5.1) and shall meet the requirement of Eq. (3.9.5.2).

$$\sigma_{\mathrm{z}} = \frac{p(1+\delta)}{(B+2h\tan\theta)(L+2h\tan\theta)} + \gamma h \tag{3.9.5.1}$$

where:

σ_{z}—stress on the surface of geotechnical plastic foam bulk (kPa);

p—vehicle load (rear axle weight)(kN);

δ—impacting coefficient, normally 0.3;

h—thickness of pavement and reinforced concrete plate (m);

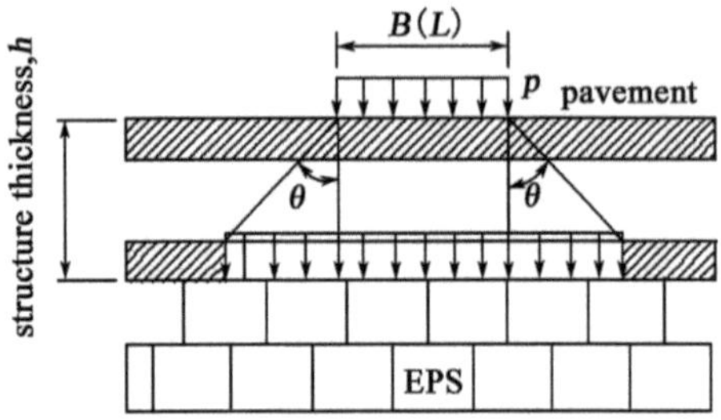

Fig. 3.3 Simplified calculation of stress distribution

B,L—width and length of contacting area of rear tire (m);

θ—load distribution angel(°),$\theta = 45°$for cement concrete pavement, $\theta = 40°$for asphalt concrete pavement;

γ—average unit weight of pavement structure and reinforced concrete plate (kN/m^3).

$$\sigma_z \le [\sigma_a] \tag{3.9.5.2}$$

where:

$[\sigma_a]$—allowable compressive strength (kPa) of geotechnical plastic foam bulk, determined by laboratory unconfined compressive test.

4 Aside from overall stability check for geotechnical plastic foam embankment design, sliding stability between geotechnical plastic foam bulk and at the bottom plate of geotechnical plastic foam embankment shall be calculated with Eq. (3.9.5.3). Safety coefficient of sliding stability shall not be less than 1.5.

$$F_h = \frac{(W + P_V)\mu + cB}{P_H} \tag{3.9.5.3}$$

where:

F_h—coefficient of sliding stability geotechnical plastic foam bulk;

W—weight of geotechnical plastic foam bulk (kN);

P_V—vertical component of resultant soil pressure (kN);

P_H—horizontal component of resultant soil pressure and earthquake transverse force (kN);

μ—friction coefficient of bottom plate and base;

c—cohesive force of bottom plate and base (kPa);

B—width of bottom plate (m).

3.9.6 Design of foamed light weight soil embankment shall be in accordance with:

1 Upright fill height of light foam soil embankment should not be greater than 15 m. The minimum fill height should not be less than 1.0 m. Bottom width of the fill body should not be less than 2.0 m when ground cross slope is big or used for embankment expansion.

2 Galvanized wire mesh and geomembrane should be used at the top of light foam soil embankment and extend to the regular embankment side for at least 2.0 m. When

the height of light foam soil exceeds 1.0 m, a galvanized wire mesh layer should be used at 0.5 m depth from the top surface.

3 Concrete block pavement may be used to protect slope when the height of upright embankment is less than 3 m; otherwise, reinforced concrete retaining wall shall be used.

4 Deformation joint shall be set along the longitudinal direction for light foam soil embankment with a suitable interval of 10–20 m. Joint should have a width of 10–20 mm and filled with foam plastic board.

5 Light foam soil beneath underground water level is only used for settlement control and may be not necessary to adopt measures to insulate underground water. When it is used for lad reduction of underground structure or underground pipeline, measures to cut off and drainage water should be adopted.

6 Performance index value for light foam soil at different environment and engineering conditions shall be in accordance with specifications in Table 3.22 for the calculation of light foam soil embankment design.

7 Overall settlement adjustment factor should be 1.0–1.1 for subgrade settlement calculation and takes low value when load bearing capacity of subgrade exceeds twice of the embankment load.

8 Aside from overall stability calculation, light soil embankment shall carry out anti-sliding and anti-overturning check when the bottom of embankment has inclined surface at the, or when the height ratio of light foam soil filled zone exceeds 1 and the height exceeds 3 m. The check shall be in accordance with the specifications in Clause 5.4.2.

9 When used as replacement above the underground structure or pipelines to reduce the weight, the integral weight of light soil and other loads shall not exceed 90% of the maximum allowable load of the underground structure or pipelines.

3.9.7 Design of fly ash embankment shall be in accordance with:

1 Fly ash embankment may be constructed entirely with fly-ash (pure fly ash) or partially with fly-ash (alternated layer of fly ash and soil). Soil slope protection shall be used for slope and shoulder. Upper roadbed zone shall adopt soil fill, or may be in combination with pavement structure, use materials of subbase course such as lime soil, lime fly ash soil to pave capping layer.

2 When soil slope is used, the requirement for setting a draining seepage ditch in the soil slope shall be determined according to construction season and local rainfall. Inverted filter shall be set outside of the draining seepage ditch.

3 The bottom of fly ash embankment shall be placed 500 mm higher than the level of underground water or long-term surface water. Otherwise, an isolation layer shall be set. The thickness of the isolation layer should not be less than 300 mm and its slope should not be less than 3%.

4 Compaction degree standard of fly ash embankment shall be determined by test on the basis of specifications in Tables 3.3 and 3.5.

5 If embankment is higher than 5.0 m, its self-stability shall be checked and calculated. The stability safety factor against sliding shall meet the requirements specified in Table 3.11.

Table 3.22 Performance index value for design calculation

Check content	Index for check	Index value for check	
		Underground water condition	Index value
Self-weight stress calculation for settlement check	Unit weight of light soil R (kN/m^3)	Above underground water	Wet unit weight of construction R_{fw}
		Below underground water	$R = (1.1–1.3)R_{fw}$
Self-weight stress calculation for load on structure check	Unit weight of light soil R (kN/m^3)	Above underground water	Wet unit weight of construction R_{fw}
		Below underground water	$R = (1.1–1.3)R_{fw}$
Self-weight stress calculation for floating resistance check	Unit weight of light soil R (kN/m^3)	Above or Below underground water	Wet unit weight of construction R_{fw}
Overall stability check of embankment	Cohesive force c and internal friction angle φ of light soil	Above underground water	Determined by test; otherwise, $c = 120$ kPa, $\varphi = 6°$
		Below underground water	Determined by test; otherwise, $c = 100$ kPa, $\varphi = 4°$
Check of anti-sliding and anti-overturning stability	Friction coefficient with contacting surface of pebble soil, sand soil and base rock	Above underground water	0.6
		Below underground water	0.5
	Friction coefficient with contacting surface of cohesive soil and strong weathering layer	Above underground water	0.5
		Below underground water	0.4

3.10 Embankment with Industrial Waste Material

3.10.1 When industrial waste material is used in embankment, it must be in accordance with the national environmental protection specifications. Industrial waste material containing hazardous materials must not be used as embankment filler.

3.10.2 Industrial waste materials that may be used for embankment, such as blast furnace slag, steel slag, and coal gangue, shall be in accordance with:

1 Blast furnace slag and steel slag shall have stable decomposition, particle size in line with requirement, and adequate strength. Immersion expansion ratio shall not be greater than 2.0%. Autoclave pulverization rate shall not be greater than 5.0%. Metallic iron content in slag shall not be greater than 2.0%. Free calcium oxide content shall not be greater than 3.0%. Slag shall be stockpiled for more than 1 year before use.

2 Coal gangue that is not fully oxidized and aged, or of plasticity index greater than 10 should not be directly used to fill embankment in Motorway and Class-1 highway. Inferior performance coal gangue shall be used through modification and test demonstration.

3 Ingredients of SiO_2、Al_2O_3 and Fe_2O_3 shall not be less 70% mass of coal gangue and the ignition loss shall not be greater than 20%. Coal gangue should not contain impurity.

3.10.3 Before used in embankment, other industrial waste materials shall be demonstrated by test and approved by administrative department.

3.10.4 Industrial waste materials shall not be used in soaked section and flooded section.

3.10.5 A comprehensive design of industrial waste material embankment cross-section profile, structure, waterproof and drainage system, and protection facilities shall be carried out in accordance with the environment condition, industrial waste material property, filling zone and so forth. The designed industrial waste material embankment shall have adequate strength and stability and prevent the contamination of industrial waste material to surface water, underground water and soil etc.

3.10.6 The following evaluation work shall be performed in the industrial waste material embankment design:

1 Analysis test of chemical ingredient and mineral ingredient is performed to identify the chemical ingredient, mineral ingredient, hazardous substance content in leaching liquid, pH value, and ignition loss content etc. to evaluate its impact on water, soil and so on. The test shall be in accordance with the specifications in *Solid Waste-Extraction Procedure for Leaching Toxicity* (GB/T 15,555).

2 Test of autoclave pulverization rate and immersion expansion ratio is performed to evaluate the stability of steel slag. The test shall be in accordance with the specifications in *Test Method for Stability of Steel Slag* (GB/T 24,175).

3 Compaction test is performed to identify maximum dry density and optimum water content.

4 Internal friction angle φ, cohesive force c, compressive coefficient, expansive coefficient, resilient modules and CBR value shall be determined by tests.

3.10.7 Structure design of industrial waste material embankment shall be in accordance with:

1 Close form embankment structure shall be adopted for industrial waste material embankment. Soil slope protection measures shall be taken for slope and shoulder. A draining seepage ditch in the soil slope shall be set. Inverted filter shall be set outside of the draining seepage ditch.

2 For industrial waste material embankment, upper roadbed zone shall adopt soil fill, or may be in combination with pavement structure, use inorganic binding stabilized soil of subbase course to pave capping layer.

3 The bottom of industrial waste material embankment shall be placed 500 mm higher than the level of underground water or long-term surface water, and an isolation layer shall be set. The thickness of the isolation layer should not be less than 500 mm. Isolation layer material may use cohesive soil of plasticity index no less than 6 and adequate strength.

4 If embankment is higher than 4.0 m, an soil interlayer may be set in the middle of substructure.

3.10.8 If embankment is higher than 5.0 m, its self-stability shall be checked and calculated. The stability calculation method and anti-sliding coefficient shall meet the requirements specified in Sect. 3.6.

3.10.9 Compaction degree standard of fly ash embankment shall be determined by test on the basis of specifications in Table 3.5.

3.11 Borrow Pits and Spoil Sites Substructure

3.11.1 Borrow pits and spoil ground shall be placed coordinately designed in accordance with the required borrow or spoil soil quantities of each section and in combination with drainage, topography, soil conditions, construction methods, land saving and environmental protection etc.

3.11.2 The allocation of borrow pits shall meet the following requirements:

1 The distance between the borrow pits and roadbed need to be far enough to prevent affecting the slope stability.

2 The borrow pit should not be located at both sides of the abutment approach.

3 The borrow pit that is to be used as a drainage pit shall ensure an easy and smooth drainage. The depth of the pit should not exceed the local underground water level but shall be adapted to the inlet height of bridge and culvert. Its longitudinal slope shall not be less than 0.2%; in flat sections the longitudinal slope shall not be less than 0.1%.

3.11.3 The allocation of spoil grounds shall meet the following requirements:

1 The distance between the spoil grounds and roadbed shall be far enough to prevent affecting the slope stability.

2 If spoil is deserted along a river, it shall prevent the aggravating the erosion of downstream roadbed and river bank while avoiding blockage and pollution to the river. When necessary, protective support and retaining structures shall be set.

3.11.4 Roadbed spoil shall be regularly stacked and properly compacted. Necessary measures of drainage, protection and greening shall be provided to assure slope stability and prevent water and soil loss.

Chapter 4
Substructure Drainage

4.1 General Requirements

4.1.1 Highway substructure waterproof and drainage design shall be carried out in accordance with the meteorology, hydrology, terrain, geology and set of bridge, culvert and tunnel. The design shall follow the principles of master-planning, rational distribution, less occupation of agricultural land and environmental protection and set perfect, unobstructed drainage system. A comprehensive design of substructure waterproof and drainage, substructure treatment and protection etc. shall be conducted, which shall be in coordination with the waterproof and drainage systems of pavement, bridge, culvert, and tunnel etc.

4.1.2 Surface water of road boundary should not flow into the bridge, tunnel and their drainage systems.

4.1.3 For sections of low-fill, shallow excavation, or drain difficulties, the measures to prevent, drain and intercept water shall be adopted to intercept the water that can probably enter into road boundary, drain the motorway out of substructure, and isolate underground water. The substructure shall maintain a dry or medium wet state.

4.1.4 Waterproof and drainage design of substructure along the river side shall arrange the layout of drainage facilities and treat the outlet of drainage facility reasonably in accordance with the river hydrological characteristics, design flood level, discharge and topography geological conditions of river course. The design shall be in coordination with river diversion facilities and buildings to prevent water erosion of substructure slope and river bank.

4.1.5 Various types of drainage facilities shall be designed to meet the performance requirements, ensure the safety of the structures and be convenient for construction, inspection and maintenance. The strength of materials used in drainage facilities shall be not less than the requirements of Appendix Table G.1.

4.1.6 Substructure drainage design shall be coordinated with the irrigation and drainage system of farmland.

W. Wu, *Specifications for Design of Highway Substructure*, JTG Standards and Specifications, https://doi.org/10.1007/978-981-96-2313-6_4

4.1.7 Temporary drainage facilities for the construction site should be combined with the permanent drainage facilities.

4.2 Surface Drainage

4.2.1 The rainfall return period for designing substructure surface drainage facilities shall be in compliance with the following requirements: Motorway and Class-1 highway is 15 years, any other classified highway is 10 years. The cross-section dimensions of various types of surface drainage facilities shall meet the requirements of designed flow volume of the drainage. The top elevation of the drainage ditch shall be 0.2 m higher than the designed water level inside the ditch.

4.2.2 Surface drainage facilities of substructure consist of side ditch, intercepting ditch, drainage ditch, drop water and chute, evaporation pond, oil–water separation pond, drainage pumping station and so on. They shall be arranged in combination with the topography and natural waterway system. The locations of inlets and outlets shall be chosen and reasonably treated to prevent blockage, overflow, leakage, sedimentation, erosion, freezing or any other problems.

4.2.3 Drainage design of the substructure in the sensitive area of water environment shall take necessary measures to protect water environment.

4.2.4 Side ditch design shall be in accordance with:

1 The types and dimensions of the side ditch cross-section shall be determined in accordance with the intensity of rainfall, catchment area, topographic and geological conditions and the influence on road side safety and environmental landscape. When conditions permit, it should use the triangular or shallow dish shaped side ditch.

2 The longitudinal slope of the side ditch bottom should be identical to the route longitudinal slope but should not be less than 0.3%. Under difficult conditions, it may be reduced to 0.1%.

3 When the side ditch erosion intensity is greater than the permissible velocity specified in Table 4.1, necessary protection and reinforcement measures shall be taken.

Table 4.1 The maximum allowable velocity of open trench

Open trench classification	Maximum allowable velocity (m/s)	Open trench classification	Maximum allowable velocity(m/s)
Fine grained soil sand	0.8	Slice rubble(pebble gravel) reinforcement	2.0
Low liquid limit silt, low liquid limit clay	1.0	Dry-laid rubble	2.0
High liquid limit clay	1.2	Mortar-laid rubble	3.0
Turf revetment	1.6	Cement concrete	4.0

4 Highway, rectangle ditch of excavation section of Class-1 highway should add reinforced concrete cover plate with drain holes or roadside guardrail. The strength and thickness of reinforced concrete cover plate shall meet the requirements of vehicle load.

5 In the seasonal frost area, the dark buried pipe (groove) under the shallow dish shaped side ditch shall be set under the maximum substructure frost depth line. Buried pipe (ditch) outlet shall take insulation measures.

4.2.5 Intercepting ditch shall be in accordance with:

1 Intercepting ditch shall be set in accordance with topographic conditions, catchment area and so forth. The topside drainage ditches of the cutting shall be set 5 m outside the top slope edge and should be combined with topography. The distance from the intercepting ditches on the upside from embankment to the toe of the embankment slope shall not be less than 2 m.

2 The types of drainage ditch cross-sections shall be determined through the conditions of their locations, discharges, topography and slopes. The longitudinal slope of the ditch bottom should not be less than 0.3%.

3 The water flow of intercepting ditches shall be discharged beyond the road boundary and should not be channeled to ditches in cutting sections.

4 The intercepting ditch shall be reinforced by anti-seepage measures.

4.2.6 Drainage ditch shall be designed in accordance with:

1 Drainage ditch shall be used to drain water out of side ditches, intercepting ditches, borrow (spoil) pits and depressions near the substructure.

2 Determination of the cross-section of drainage ditch shall be combined with the topographic and geologic conditions. The longitudinal slope of ditch bottom should not be less than 0.3%. The connection between the drainage ditch and other drainage facilities shall be smooth. If the drainage ditch is vulnerable to water erosion, measures of protection and reinforcement shall be taken in accordance with the actual situation.

4.2.7 Drop water and torrent chute should be used when water flow through the steep slope section with a slope more than 10% or water head difference of more than 1.0 m, or in special steep area. The design of drop water and torrent chute shall be in accordance with:

1 Reinforcement measures shall be taken for drop water and torrent chute.

2 The longitudinal slope at the bottom of the torrent chute shall be fit with the topography. The inlet shall be protected and strengthened, and in order to prevent erosion, energy dissipation measures shall be taken at the outlet.

3 To prevent base sliding, anti-sliding flatbeds or tenons shall be set at the bottom of the torrent chute.

4.2.8 For sections with dry climate and being difficulty to drain, surface water may be gathered in borrow pits along the road or in evaporation ponds which are set specially. Evaporation pond shall be designed in accordance with:

1 The principle is to keep an appropriate distance from the edge of the evaporation pond to the outer edge of the side ditch to ensure the stability and safety of the roadbed. The distance shall not be less than 5 m, and it shall not be less than the collapsible

radius in a collapsible loess area. The design water level of the evaporation pond shall be lower than the bottom of drainage ditch.

2 The basis for the design of the evaporation pond shall be that the total volume of rainwater flowing into the pond from the roadbed in a month can be timely infiltrated and evaporated. The capacity of each evaporation pond shall be determined by hydraulic and hydrographic calculation to prevent salinization or swamping.

3 Appropriate protections and strengthening measures shall be taken in accordance with the specific situation of the evaporation pond.

4.2.9 If the outlet of roadbed drainage ditch is located in an area where the water quality is particularly sensitive, an oil–water separation pond should be set. The quality of the discharged water shall meet the provisions in *Integrated Wastewater Discharge Standard* (GB 8978). Oil–water separation pond shall be in accordance with:

1 Precipitation methods should be adopted in oil–water separation. Wastewater shall be channeled through mesh screens and desilting basins before entering into the oil–water separation pond.

2 The dimensions of the oil–water separation pond shall be determined by water amount collected from drainage ditches in the corresponding road section and shall ensure that the oil–water flown into the pond has enough time to finish the separation, filtration or purification.

4.2.10 Down channel shall be in accordance with:

1 An independent and perfect drainage system shall be set up in the down channel to eliminate surface runoff of catchment areas and groundwater that influences the road function. Drainage facilities shall be coordinated with other drainage facilities around.

2 For surface drainage runoff calculation, the design return period should not be less than 5 years.

3 Gravity flow drainage should be set for down channel. Drainage method may be determined according to Table 4.2 when the drainage condition is restricted.

Table 4.2 Drainage Methods of Down Channel

Drainage Method	Applicable Conditions
Gravity Drainage	May be applicable to the bottom of channel is higher than the normal water level of the bottom of canal
Pump Station Drainage	May be applicable to channel with high precipitation, high groundwater level and lower bottom than that of canal can not be self-discharged
Seepage Well Drainage	May be applicable to channel drainage with annual precipitation less than 600 mm, low groundwater level, good aquifer permeability and depth less than 10 m. The water flow in the channel shall be filtered through the oil–water separation tank before entering the seepage well to protect the groundwater quality
Evaporate Pond Drainage	May be applicable to channel drainage with annual precipitation less than 400 mm, high evaporation intensity and low groundwater level

4.2.11 If the water catched in the roadbed area cannot be discharged freely, the drainage pumping station may be set up. The drainage pumping station includes the collecting tank and pump house. Drainage pumping station shall be in accordance with:

1 The volume of the collecting tank shall be determined by factors such as collected water amount and the capacity and working condition of the pump, etc.

2 The water pumped out of the collecting tank shall be discharged outside the road boundary.

3 Drainage outlet, drainage ditch and other drainage facilities shall be set at both ends of the excavation section to intercept and induced channel the surface water from downstream.

4 The other design aspects of drainage pumping station shall be in accordance with the provisions in *Design Code for Pumping Station* (GB 50,265).

4.2.12 Design of substructure drainage in interchange area shall be in accordance with:

1 The design of substructure drainage in interchange area shall set good and smooth drainage system.

2 The type and structure of substructure drainage in interchange area shall be determined in accordance with interchange type, catchment area, terrain and geology, climate and environment landscape etc. Substructure drainage facilities on the ramp and mainline shall be connected.

4.2.13 Waterproof and drainage design of the central separation belt shall be in accordance with:

1 When the surface of central separation belt is paved, it shall adopt inclined cross slope at both sides. The slope of the central separation belt should be identical to that of pavement. Waterproof layer shall be placed beneath the paved central separation belt.

2 When the surface of central separation belt is unpaved, it shall adopt inclined cross slope at both sides, a comprehensive waterproof and drainage system of waterproof layer, longitudinal drainage sewer, water tank and transverse drainage pipe etc. shall be implemented inside the central separation belt. Sewer should be set under communication pipe.

3 The surface of the concave central separation belt should be set to a shallow dish with a suitable slope of 1:4–1:6. A comprehensive drainage system of longitudinal side ditch, water collecting well, horizontal drainage pipe, side slope jet, and energy dissipation pool etc. shall be set in the central separation belt. The section size and spacing shall be determined by the hydraulic calculation.

4 The waterproof layer shall be set between the backfill soil of central separation belt and the pavement structure.

4.3 Underground Drainage

4.3.1 Before the design of underground drainage, the field engineering geological and hydrogeological surveys, explorations and tests shall be carried out to find out hydrogeological conditions and obtain related hydrogeological parameters.

4.3.2 When the underground water influence the substructure stability or strength, measures to intercept, channel, drain, lower or separate water shall be adopted in accordance with the type of groundwater, buried depth of aquifer, permeability of stratum and its influence on environment. The underground drainage facilities shall be coordinated with the surface drainage facilities. The form of underground drainage facilities may be determined in accordance with:

1 When the groundwater is shallow or without fixed aquifer, the isolation layer, drainage cushion, blind ditch, and sewer etc. may be used.

2When the groundwater is deep or without fixed aquifer, the upward-inclinedly drilled drainage hole, seepage well and drainage tunnel etc. may be used.

4.3.3 Drainage cushion and isolation layer shall be in accordance with:

1 When the underground water level of clay soil section is less than 0.5 m or the underground water level of silty soil section is less than 1.0 m, drainage cushion layer and isolation layer should be set at the bottom of the low embankment filled with fine grained soil.

2 Drainage cushion thickness shall not be less than 0.3 m. Cushion material should use natural gravel or middle-sized coarse sand. When the composite waterproof and drainage plate is used as the isolation layer, the drainage cushion layer may not be arranged.

3 The isolating layer may use geotechnical synthetic materials such as geotechnical membrane, composite geotechnical membrane and composite waterproof board etc. The thickness, texture and type of impervious material shall be determined according to the climate and geological conditions. Geotechnical synthetic materials shall be in accordance with the *Technical Specifications for Application of Geosynthetics in Highway* (JTG/T D32).

4.3.4 Blind ditch (pipe) shall be in accordance with:

1 The blind ditch (pipe) may be used to discharge spring water or underground concentrated flow.

2 The longitudinal slope of the blind ditch bottom should not be less than 1.0% and should not be less than 0.5% even in difficult conditions. The longitudinal slope near the outlet shall be increased and the outlet shall be 0.2 m higher than the normal water level of the surface drainage ditch.

3 Blind ditch may be constructed by mortar rubble or precast concrete block masonry. The blind ditch roof shall be set with concrete or stone cover plate. The thickness of the soil on the top surface of the cover plate shall not be less than 0.5 m. Section size of blind ditch shall be determined according to the water flow and the terrain and geological conditions.

4.3.5 Underdrain shall be in accordance with:

1 Underdrain shall be in sections of excavated substructure with groundwater exposure, inclined embankment, cut-fill transitional substructure, and low embankment with groundwater depth less than 0.5 m.

2 Underdrain type shall be determined in accordance with the groundwater storage condition, seepage flow, location and drainage distance, as specified in Table 4.3. The underdrain section size shall be calculated by the groundwater seepage flow.

3 The burial depth of the underdrain shall be determined by factors such as the elevation and depth to be lowered of the underground water, and permeability coefficient of the aquifer medium and so on. The base of intercepting seepage drain should be set no less than 0.5 m down to the water-resisting layer. The base of the revetment underdrain or slope underdrain should be set on solid soil under the aquifer.

4 The minimum longitudinal slope of rock-fill underdrain and no-fines concrete underdrain should not be less than 1.0%, and for the pipe underdrain and tunnel underdrain, it should not be less than 0.5%. The outlet shall be 0.2 m higher than the normal water level of the surface drainage ditch.

5 Slope underdrain and revetment underdrain shall be vertically embedded in the slope body and arranged in the form of strip bifurcation, or arch shape, and its interval should be 6–10 m.

6 Underdrain materials shall use clean gravel, coarse-sized sand, crushed stone and rubble whose fine material content less than 2.36 mm shall not be greater than 5%. Underdrain wall shall be set with permeable geotextile or medium coarse-sized sand filter layer. Seepage pipe may use HPPE with holes, PVC, PE pipe, soft permeable pipe, and non-fines concrete pipe etc.

4.3.6 Upward-inclinedly drilled drainage holes shall be in accordance with:

1 Upward-inclinedly drilled drainage holes (UDDH) may be used for the discharge of groundwater within the slope.

2 The elevation angle of UDDH should not be less than 6° and the length of UDDH shall extend to a groundwater-enriched zone or to the potential sliding surface. UDDH should be distributed in groups in accordance with the water seepage conditions of the slope.

3 Inlet of UDDH and seepage pipe section shall be wrapped with permeable geotextile to prevent blockage of seepage holes.

4.3.7 Seepage well shall be in accordance with:

Table 4.3 Applicable conditions of various underdrains

Underdrain classification	Applicable conditions
Rock-fill underdrain, non-fines concrete underdrain	May be used in sections where the groundwater flow is not large and the drainage distance is short
Pipe underdrain	May be used in sections where the groundwater flow is large, the groundwater level is shallow and the drainage distance is short
Tunnel underdrain	May be used in sections where the groundwater flow is large and the burial depth is deep

1 Seepage well may be used to intercept and lead the deep underground water with fixed aquifer, and to drain the surface water of the digging channel.

2 A multiple-well layout should be used for the interception and groundwater drainage wells and cooperates with the drainage tunnel and other drainage facilities. The direction of seepage well arrangement should be perpendicular to the direction of seepage, and its depth should pass through the aquifer. The section size and spacing shall be determined by seepage calculation. Clean sand and gravel should be used inside the seepage well, and the filter layer shall be arranged between the well wall and the filling material.

3 When it is used for surface water drainage of the digging channel, the seepage well should be 10 m away from embankment slope toe. The seepage well size shall be determined according to the water flow of digging channel by hydraulic calculation. Reinforced concrete pipe or corrugated pipe should be used for seepage well. The upper part is the water collection well and the lower part is the seepage well. Seepage well materials shall use clean gravel and rubble gravel etc. whose fine material content less than 2.36 mm shall not be greater than 5%. Seepage well wall shall be set with anti-filter layer.

4.3.8 Drainage tunnel shall be in accordance with:

1 Drainage tunnel may be used to intercept and lead the deep underground water. The burial depth shall be determined by the burial depth of the main aquifers, and shall be embedded in the stable strata. The top of drainage tunnel shall be below the sliding surface (zone) not less than 0.5 m.

2 For the other aquifers above the sliding surface, seepage well or pipe should be set on the top of seepage tunnel to channel the water into the hole. For the confined aquifer under the seepage tunnel, seepage holes should be set in the bottom of the hole.

3 Tunnel section size shall be determined by the groundwater inflow calculation. The clearance height should not be less than 1.8 m and the clearance width should not be less than 1.0 m.

4 The tunnel plane axis should be smooth and straight, and the bottom slope shall not be less than 0.5%. Different slope segments may be connected by the broken line slope or the stepped drop water etc.

5 Tunnel structure design shall be in accordance with the provisions in *Code for Design of Road Tunnel* (JTG D70).

4.3.9 Inspection and clear well shall be in accordance with:

1 The inspection and clear well should be set in the straight line segment for certain distance, the plane turning point, the grade change point of longitudinal slope and so on for the long and deep ditch (canal), underdrain and seepage tunnel.

2 The inspection well shall be equipped with inspection ladder and wellhead cover. When the inspection well is also used as seepage well, a filter layer shall be set along the well all.

4.3.10 In the frost vulnerable area of complicated hydrogeological condition, the drainage tube of underdrain shall be set below the roadbed freezing depth for no less than 0.25 m. For the underdrain and seepage tunnel in the heavy frost zone, anti-freeze measures shall be taken.

Chapter 5
Substructure Protection and Retaining Structures

5.1 General Requirements

5.1.1 Comprehensive measures of combining engineering protection with flora protection shall be taken for all classes of highways in accordance with the local climate, hydrology, topography, geological conditions and the distribution of road materials to control substructure diseases, ensure substructure stability, and be in harmony with the surrounding environments and landscapes.

5.1.2 Substructure slope protection works shall be set on stable side slopes. When the soil and climate is suitable, plant protection should to be used; when the slope protected by plant would be scoured, mortar rubble or concrete aggregate shall be set; for complete stable slope consisted of partly weathered, slightly weathered, unweathered hard rock, protection may be spared. In sections with unstable Substructure, an integrated design of slope protection and reinforcement shall be stressed.

5.1.3 The engineering geological survey of slope and foundation shall be conducted when designing reinforcement. The engineering geological and hydrogeological conditions and its potential for corrosion, adverse geology and special geotechnical distribution as well as foundation bearing capacity and anchoring conditions shall be found out. Physical and mechanical parameters of rock and soil mass properly need to be determined.

5.1.4 The design of a substructure retaining structure shall ensure stability, solidity, and durability of Substructure in various design load combinations. The determination of structural types and locations shall ensure safety and reliability, economy and rationality and easy construction and maintenance. Structural materials shall meet the requirements of durability and corrosion resistance.

5.1.5 The design of retaining structures shall coordinate with bridge abutments, tunnel portals and existing retaining structures with smooth connections.

5.1.6 In developed groundwater sections, the integrated design for substructure slope protection and underground drainage measures shall be carried out. In rainy regions, comprehensive measures of slope protection, water interception

W. Wu, *Specifications for Design of Highway Substructure*, JTG Standards and Specifications, https://doi.org/10.1007/978-981-96-2313-6_5

and drainage shall be taken to protect the embankment constructed by sandy soil, fine-grained soil.

5.1.7 Strength of retaining structure material shall be not less than requirements in appendix Table G.2. Other materials shall meet the provisions of current national standards.

5.1.8 Temporary slope protection measures shall be made in substructure construction, which should be combined with permanent protection engineering.

5.2 Slope Protection

5.2.1 For slope which is vulnerable to natural damage, protective measure shall be chosen after technical and economic comparison in Table 5.1 based on climate, rock properties, height, hydrogeological condition, construction condition and conservation requirement and environmental protection.

5.2.2 Plant protection should be combined with grass/shrub/tree and shall choose local dominant populations. They shall comply with the following requirements:

1 The minimum thickness of soil layer for grass shall not be less than 0.15 m, for shrub shall not be less than 0.30 m.

2 The thickness for spray-sowing should not be less than 0.10 m. The material percentage of soil, grass fiber, slow-release fertilizer nutrition, glue as well as water retention agent etc. shall be determined with tests.

5.2.3 Anchored concrete frame, polygonal hollow concrete block, square mortar rubble or cement aggregate may be adopted for frame plant protection. Planting soil may be filled in frame for the seed spraying and grass planting. For areas with high rainfall intensity, frames should be made as the types of the intercepting ditch. Anchored concrete frame may be applicable to rock cut slopes that are weathered and cracked.

5.2.4 Shotcrete and hanging net shotcrete shall meet the following requirements:

1 Mortar or cement concrete may be used for shotcrete. The thickness of shotcrete support for the side slope should not be less than 50 mm. The thickness of sprayed concrete protection should not be less than 80 mm.

2 The thickness of anchor hanging net shotcrete shall be less than 0.25 m and more than 0.10 m. The thickness of reinforced protection layer shall not be less than 20 mm.

3 On the shotcrete supported slope, outlets and expansion joints shall be set.

4 Climbing plants shall be set in shattered and slope platform.

5.2.5 Slope protection design shall meet the following requirements:

1 The thickness of dry-laid rubble slope protection should not be less than 0.25 m.

2 The thickness of the mortared rubble slope protection should not be less than 0.25 m and expansion joints and outlets shall be set in the slope protection.

3 Blinding layer of crushed stone or sand & gravel shall be set under the paving layer, and the thickness should not be less than 0.10 m.

Table 5.1 Types of slope protection and application conditions

Type of protection	Subclass	Application situation
Plant protection	Vegetation protection	May be applicable to the soil side slope with a slope ratio less than 1:1. When the slope is too high, vegetation protection may be combined with geonets
	Turf	May be application to the soil side slope with a slope ratio less than 1:1 or totally/highly/weak weathered rock slope
	Tree planting	May be applicable to the protection of soil, soft rock and weathered rock slope with slope rates less than 1:0.75
	Spray-sowing	May be applicable to the protection of sandy soil, soil-rock, coarse grained soil, giant grain soil and weathered rock slope with slope rates less than 1:0.75 and its height should be less than 10 m
Frame plant protection	–	May be applicable for soil slopes with the slope rate less than 1:0.75 and completely weathered rock slopes
Masonry protection	Shotcrete support	May be applicable to the rock slopes with a slope rate of less than 1:0.5, which are susceptible to be but not highly weathered yet. Should not be allowed in highway and roads with high requirement for environment
	Hanging nets	May be applicable to the rock slopes with a slope rate of less than 1:0.5, which are susceptible to be weathered. Should not be allowed in highway and roads with high requirement for environment
	Dry-laid rubble slope protection	May be applicable to soil (rock) cut slopes with a slope rate less than 1:1.25
	Mortared rubble slope protection	May be applicable to soil cut slopes and rock cut slopes of which the slope ratio is less than 1:1 and the rocks are susceptible to getting weathered
	Facing wall	May be applicable to soil cut slopes and susceptible to getting weathered rock cut slopes which the slope ratio is less than 1:0.5

5.2.6 The height of the single-stage facing wall should not exceed 10 m and expansion joints and outlets shall be set. The facing wall shall be constructed on a solid base. In a frost region, the foundation shall be set on less than 0.25 m under the frozen depth. The front toe of the facing wall shall be lower than the bottom surface of side ditches.

5.3 Substructure Protection Along Riverside

5.3.1 When the substructure along a river is flushed by water, appropriate types of protection works, river diversion works or river course change projects shall be adopted based on Table 5.2 according to the location of the substructure and other factors such as features of the river, properties of the flow, geomorphology and geology of the river course etc.

5.3.2 The top elevation of scour protection works shall be the designed water level plus the hitting wave height, backwater height and safety height. The foundation of scour protection works shall be buried no less than 1 m below the scour depth or should be embedded within the bed rocks. Pile foundation, or suitable surface protection may be adopted when the scour is deep and underwater construction is difficult.

Table 5.2 Type of protection and application situation

Type of protection		Application situation
Plant protection		May be applied to the seasonal water scour where the permissible velocity of flow is less than 1.2 m/s to 1.8 m/s and the direction of flow is parallel to road without a mainstream scour of the flood. Grass protection should not be used for the embankment slopes which are frequently or chronically immersed in water
Masonry or concrete slope protection		May be applied to embankment slopes with a permissible velocity of flow of 2 m/s to 8 m/s
Soft sinking mattress of geotextile and geofabriform		May be applied to the substructure scour protection along the river with the permissible velocity of flow of 2 m/s to 3 m/s
Rock filled gabion		May be applied to the substructure along the river and embankment with the permissible velocity of flow of 4 m/s to 5 m/s
Water immersion retaining wall		May be applied to the canyon current with a permissible velocity of flow is 5 m/s to 8 m/s and river section with severe scour
Apron		May be applied to a road section where scour against the retaining wall along the river, or part of the slope is intensively scoured and the deep foundation construction for that is inconvenient
Riprap		May be applied to the foundation protection of retaining wall, slope protection, substructure side slope or the slope toe which is often immersed deeply in water
Pile protection		May be applied to river bend with too much local scour depth or wide-shallow river
River diversion	Groin	May be applied to river sections in a wide-shallow change state to jet flow or to decrease the velocity of flow, so as to reduce the scour to the river bank and substructure
	Longitudinal dam	May be applied to river banks with narrow river cross sections and poor geological conditions. It also provides substructure protection along the river so as to adjust the curvature of flowing water and improve flow pattern

5.3.3 Protection shall be connected smoothly with upstream and downstream bank slope. The depth of end into the bank shall be enough to prevent the deterioration of upstream and downstream hydrological conditions.

5.3.4 In the process of setting a water division structure, a water diversion route shall be designed according to the river geomorphology, geologic and flow properties, the evolution of river course and protection requirements. An aggravated scour of farmlands, villages, highway, and downstream substructure shall be avoided. In the sections of valleys in mountainous regions, regulation structures should not be set.

5.3.5 The thickness of masonry or concrete slope protection shall be determined according to the flow velocity and size of waves. The thickness of dry-laid rubble slope should not be less than 0.25 m. The thickness of mortared rubble slope shall not be less than 0.35 m. The thickness of cement concrete shall not be less than 0.10 m. Revetment underside filter layer shall be set up.

5.3.6 The design of the immersed retaining wall shall meet the relevant requirements of Sect. 5.4 of the Specifications, the checking erosion resistance shall be done. Necessary attention shall be paid to the connection between the water immersion retaining wall and bank slope.

5.3.7 The design of groin shall meet the following requirements:

1 The length of the groin shall be determined according to the protection length, intersection angle between groin and flow, terrain of river section, hydrological conditions, as well as geological conditions of the river bed, etc. The projected length, which is perpendicular to the direction of the flow, should not exceed 1/4 of the width of the stable river bed.

2 Overflow dam or submerged dam should be adopted for the groin protection of the substructure. The intersection angle between the groin and flow direction should be less than or equal to 90°.

3 When group dams are set, the distance between the dams shall not be larger than the protection length of the front dam. When the permissible velocity of flow that the bank or substructure between the groins can bear is less than the velocity of the back flow, the distance between the dams shall be shortened or protective measures shall be taken for the bank and substructure slope.

4 The cross-sectional form and dimensions of groin shall be determined according to material type, hydrological characteristics of the river, etc. The width of the dam crest shall be determined by the stability calculation.

5.3.8 The design of longitudinal dam shall meet the following requirements:

1 The connection between the longitudinal dam and upstream and downstream river banks shall ensure to smoothen the water flow. The starting point shall be chosen at the transition section where the flow is neat and smooth and the dam root should be located above the turning point of the main flow.

2 The width of the dam crest shall be determined by stability calculations, and the dam foundation shall be embedded not less than 3 m within the stable banks.

3 For the longitudinal overflowing dam, a checker dam shall be set behind it.

5.3.9 The design of river diversion shall comply with the following requirements:

1 If the substructure along the river is seriously scoured by water flow, or it is hard to make protection works, or there are many crossings through tortuous river channel within short distances, the river may be diverted.

2 Diversion should not be applied the rivers with main channels that change frequently or have too many tributaries.

3 The locations of the starting point and end point shall be smoothly connected with the original river bed. To prevent the water from returning to the old course, the longitudinal slope should be increased and a retaining dam or longitudinal dam should be set at the entrance of the river. The section of the new river channel shall be designed based on the discharge of designed flood frequency.

5.4 Retaining Wall

5.4.1 The selection of types of retaining wall shall be comprehensively in consideration of factors such as geology, hydrogeology, scour depth, loading condition, environmental condition, construction condition, engineering cost etc. and as stipulated in Table 5.3.

5.4.2 The design of retaining wall shall be carried out in such a way that is dominated by partial factor method under limit state. Vehicle load shall be calculated by additional load intensity method. In a retaining wall design, bearing capacity under limit state shall be calculated and checked by limit state of normal service, as well as retaining wall stability against sliding, overturning and overall stability, which shall be in conformity with Appendix H.

5.4.3 Retaining wall should be used by open cut foundation. The embedded depth of the foundation shall meet the following requirements:

1 The embedded depth of foundation shall be not less than 1.0 m. If the retaining wall is founded on hard rock with a weathered layer that is not too thick, the bottom of the foundation shall be set below the weathered surface layer of bedrock.

2 If the foundation is subject to scour, the scour depth shall be calculated according to the design flood frequency and the bottom of foundation shall be placed not less than 1 m below the local scour line.

3 When the freezing depth is less than or equal to 1.0 m, the base shall be below the freezing line at least 0.25 m, and the minimum embedding depth shall be at least 1.0 m. When the freezing depth is greater than 1.0 m, the minimum embedding depth shall be at least 1.25 m, and countermeasures shall be taken for foundation soil 0.25 m below freezing depth to prevent frost damage.

4 The depth under road shoulder of cutting retaining wall foundation shall not be less than 1.0 m, and more than 0.2 m above ditches masonry bottom.

5 If the foundation is set on a cross slope, the embedded depth and horizontal distance between the toe of the retaining wall and the ground surface shall meet the requirements in Table 5.4. The bottom of the foundation shall be stepped if it is set on a longitudinal slope greater than 5%.

5.4.4 The design of retaining wall structure shall meet the following requirements:

Table 5.3 Applicable conditions for various types of retaining walls

Type of retaining walls	Applicable condition
Gravity retaining wall	Suitable for road shoulders, embankment and cutting constructed in normal areas, soaking area and high intensity zone. The height of the wall should not exceed 12 m, which should be limited to 6 m when the wall is dry-laid
Semi-gravity retaining wall	Suitable on the soft foundation or the foundation with high groundwater level that lack of bearing capacity to support the gravity retaining wall. The height of the wall should not exceed 8 m
Gabion retaining wall	May be applied to soil with too much underground water and weathered broken rock sections
Cantilever retaining wall	Should be applied in a filling section when stone material is scarce or the foundation lacks bearing capacity. The height of the wall should not exceed 5 m
Counterfort retaining wall	Should be applied in filling sections when stone material is scarce and the foundation lacks bearing capacity. The height of the wall should not exceed 15 m
Anchored retaining wall by tie rods	Should be applied in a cutting section where the quality of rock and the retaining body has to be set high. The wall may be used as an anti-slide retaining wall. It may be built in the form of column-anchor or as a board with single tier or multiple tiers. The height of each tier should not exceed 8 m. There shall be a platform with a width of no less than 2 m between adjacent tiers
Anchored bulkhead retaining wall	Should be applied in the road shoulder or embankment in the vicinity of which the stone material is scarce. Shall not be applied in the landslide, collapse, soft soil or expansive soil areas. The wall may be built in the forms of rib-column or plate, and its height should not exceed 10 m. The rib-column wall may be built with single or double tiers; the height of each tier should not exceed 6 m. There shall be a platform with a width of no less than 2 m between adjacent tiers; the rib columns in the upper and lower tiers should be staggered
Reinforced earth retaining wall	May be divided into the reinforced earth retaining wall with panel and geogrid reinforced earth retaining wall without panel. Panel reinforced earth retaining wall may be applied as the road shoulder or embankment retaining wall in general areas; geogrid reinforced earth retaining wall may be applied as embankment retaining wall in general areas, and shall not be applied in landslide, water scouring or collapse areas. In motorway or Class-1 highway, the height of the wall should not exceed 12 m. In Class-2 and lower highways, the height of the wall should not exceed 20 m. When using multiple tiers, the height of each tier should not exceed 10 m and there shall be a rectangular platform with a width greater than 2 m between adjacent tiers
Pile-plank retaining wall	Applied on the homogeneous rock foundation with thin surface soil or strongly weathered layer; the retaining body may be set higher. Also, it may be applied in road shoulder, embankment retaining in the earthquake area or in the landslide section

Table 5.4 Requirements for the embedment of foundation under sloped ground surface

Category of embedded layer	The minimum embedded depth h (m)	The horizontal distance from the ground surface L (m)
Common hard rock	0.60	1.50
Soft rock	1.00	2.00
Soil	≥1.00	2.50

1 The joining section between retaining wall and roadway substructure or other structures shall be designed carefully. Connection between embankment and retaining wall may use conical slope, the retaining wall shall extend into embankment structure for at least 0.75 m. The edge of the cutting retaining wall shall be embedded in the cutting slope. The embedment depth of the wall inserted into ground layer shall not be less than 1.5 m for soil, 1.0 m for weathered soft rock layer and 0.5 m for slightly weathered rock layer.

2 Water outlets shall be set inside wall body which incline wall and the slope shall not be less than 4%. Filter layer shall be set in wall back. The location and quantity of water outlets shall be arranged according to seepage of wall back. Water outlets may adapt tube type material. Filter layer shall be set in inlet, and it should be made of permeable geotextile filter layer. Filter layer in wall back should use permeable sand and gravel with the silt shall be less than 5% and the thickness shall be less than 0.5 m.

3 Expansion joints and settlement joints shall be set in monolithic wall. Expansion joints shall be set in the change section of wall body and section with other construction. Settlement joints shall be set in change places of topography and foundation. Expansion joints and settlement joints may be set together.

4 The top width of shoulder retaining wall shall not encroach on the driveway and side strip or substructure of hard shoulder. Barriers shall be set on the top.

5.4.5 Design of gravity and semi-gravity retaining wall shall meet the following requirements:

1 The top width of the retaining wall shall not be less than 0.4 m, 0.5 m or 0.6 m for the wall constructed with concrete, grouted rubble or dry-laid masonry, respectively.

2 The slope of the back of the gravity retaining wall shall be selected according to the topographic conditions and economic comparison.

3 In the design of the balance weight shoulder retaining wall, the intersection of the balance weight platform and upper wall back shall be strengthened in order to enhance the shear capacity of the cross section of wall in this intersection.

4 In the design for the semi-gravity retaining wall, the number of turning points of the stem relative to the bottom plate shall be defined to meet the requirements on flexural tensile strength and stiffness. The thickness of the top shall not be less than 0.4 m and the length of the part extended out of the front toe should not be larger than 1.5 m.

5 Retaining wall whose height is less than 10 m may use grouted rubble and whose height is more than 10 m should use concrete.

5.4.6 Gabion retaining wall shall meet the following requirements:

1 The shape of gabion retaining wall may be set as outside step, inside step and pyramid etc.

2 The gabion may be made of galvanized steel wire, galvanized iron wire and ordinary wire. For permanent project, galvanized steel wire shall be used. For projects lasting for 8–12 years, galvanized iron wire may be used. For projects lasting for 3–5 years, ordinary wire may be used.

3 Gabion shall be filled with hard rubble or block stone and the size of stone should be between 100 and 300 mm and the number of stone smaller than 100 mm shall be less than 15%. Stones shall be outside stone cage grid and the air void shall be less than 30%.

4 A layer of permeable geotextile shall be set in the gabion retaining wall to prevent siltation.

5.4.7 Design of cantilever retaining wall and counterfort retaining wall shall meet the following requirements:

1 The width of the vertical wall shall not be less than 0.2 m; the thickness of the bottom slab shall not be less than 0.3 m.

2 The section length along the counterfort retaining wall should not be larger than 20 m.

3 There should be at least 3 counterforts in each section of the counterfort retaining wall.

4 The diameter of the main reinforcement placed in the wall should not be less than 12 mm.

5.4.8 Design of anchored retaining wall by tie rods shall meet the following requirements:

1 The distance between adjacent rib columns should be 2.0–3.0 m. The rib column should be set vertically or inclined towards the fill behind the wall but the inclination slope shall not be larger than 1:0.05.

2 The platforms of multi-tier rib column-tie rod retaining walls should be sealed with a C15 concrete layer that has a thickness of no less than 0.15 m; the platform should incline to the outside of the wall with a slope of 2%.

3 The tie rods may be set in double rows or multi rows on each grade of the rib columns. The tie rods may be arranged according to the principle of equal bending moments or equal support reactions. The angle between the tie rods at each layer and horizontal should be controlled between 15° and 20°. The distance between adjacent layers of tie rods should be no less than 2.0 m.

4 The continuous reinforcement throughout the rib column entire length shall be set on the loading and opposite sides of the column. The diameter of the reinforcement shall not be less than 12 mm.

5 The vertical wall plate should have a constant thickness, and the thickness shall not be less than 0.3 m. Anchor holes on the precast wall panels shall be preset in advance.

5.4.9 Design of anchored bulkhead retaining wall shall meet the following requirements:

1 The distance between adjacent rib columns of the anchored bulkhead retaining wall should be 1.5–2.5 m. The height of each rib column level should be at 3–5 m. The rib column shall be set vertically or inclined towards the fill behind the wall. The slope of the rib column inclination should be 1:0.05. The inclination of rib column shall not be towards the front side of the retaining wall. Rib columns shall reserve round or elliptical-shaped anchor ducts for the tie rod in advance. The diameter or minor diameter of the duct shall be larger than the diameter of the tie rod.

2 The concrete foundation shall be set underneath the rib columns. The foundation type may be concrete strip foundation, isolated pad foundation, or isolated cup-base foundation. The thickness of foundation should be greater than 0.5 m, and the edge width from the column face to the edge of foundation should not be less than 0.1 m.

3 Continuous reinforcements throughout the entire length the rib column shall be placed on both faces of the column in the loading direction. The reinforcement diameter shall not be less than 12 mm.

4 The platform of the anchored bulkhead retaining wall with multi-level rib columns should be sealed with a C15 concrete layer with a thickness of not less than 0.15 m. The platform should incline to the outside of the wall with a slope of 2%. The top of the substructure should also be covered with a sealing layer if fine grained soil fill is used.

5 Each wall plate of the anchored bulkhead retaining wall need to be anchored with at least one tie rod with the diameter should be within 22–32 mm.

6 The anchored bulkhead should adopt reinforced concrete slab. For a rib column anchor plate, the slab area shall not be less than 0.5 m^2. For an anchored plate without rib columns, the slab area shall not be less than 0.2 m^2. Anchor plates need to be provided with two-way reinforcement.

7 Anti-corrosion treatments shall be done on the tie rods, connections of the tie rods, rib columns and anchored bulkheads.

5.4.10 Design of reinforced earth retaining wall with panel shall meet the following requirements:

1 The design and calculation of reinforced earth retaining wall with panel shall comply with Appendix H.

2 Reinforced earth retaining wall should use pre-casted reinforced concrete members, the thickness shall not be less than 80 mm. The horizontal alignment of reinforced earth retaining wall may be straight, broken line or curved. The angle between two adjacent walls should not be less than 70°. A concrete foundation with width not less than 0.4 m and thickness not less than 0.2 m, and embedment depth not less than 0.6 m shall be set beneath the reinforced earth retaining wall. The foundation should not be set on a longitudinal slope, it may be set on a flat or step structure according to local topography.

3 Compound earthwork geogrid or reinforced concrete plate belt should be used as brace. Earthwork geogrid shall comply with *Technical Specifications for Application of Geosynthetics in Highway* (JTG/T D32).

4 Based on the premise that the anti-pullout requirement is met, the length of the reinforcement shall be in compliance with the following requirements:

(1) When the height of the wall is larger than 3.0 m, the minimum length of the reinforcement shall be larger than 0.8 times the height of the wall and no less than 5.0 m. When unequal lengths of reinforcement are used along the wall, the height of the wall section with equal lengths of reinforcement shall be larger than 3.0 m. The length difference between the adjacent reinforcements should not be less than 1.0 m.
(2) When the height of wall is less than 3.0 m, the reinforcement shall have an equal length of no less than 3.0 m.
(3) With the use of precast reinforced concrete, the length of each section should not be greater than 2.0 m.

5 The connection between the steel strap and the wall panel shall be secure and reliable, and shall have the same corrosion resistance as the steel strap. When two sides of the retained earth are reinforced and the reinforcement of the two sides intersects with each other, the steel straps shall be staggered to avoid overlap.

6 Reinforced retaining wall should be filled with permeability coarse sand gravel or filling gravel. The contact parts between filling and reinforcement material shall not allow blocks with sharp edges. The maximum particle size of filling shall not be greater than 100 mm.

7 Waterproof or drainage measures shall be taken if there is surface water or ground water undermining the reinforced earth retaining wall. When the reinforced area is filled with fine grained soil, the width of filter layer shall be not less than 0.30 m inside wall. Measures to prevent the reinforced earth from frost heaving shall be taken in seasonal frozen areas.

8 The width of toe protection for reinforcement of slope shall be not less than 1.0 m. The depth of foundation embedment shall be calculated from the top of toe protection.

9 Concrete or reinforced concrete cap rock should be set on the top of retaining wall.

10 In a multi-level mechanically reinforced earth retaining wall, the platform between the adjacent levels shall have a transverse slope of no less than 2% at its surface for drainage and shall be protected with a concrete slab of C15 concrete and thickness no less than 0.15 m. If the fill is comprised of sandy soil, a bedding course with gravel or lime-soil with a width of no less than 1.0 m and thickness of no less than 0.5 m shall be set beneath the foundation of the upper level wall (as shown in Fig. 5.1).

5.4.11 Design of reinforced earth retaining wall without panel shall meet the following requirements:

1 When the angle between reinforced slop and horizontal plane is not be less than 70°, the design of retaining wall shall follow Appendix H. When the angle is less than 70°, the design and calculation shall comply with *Technical Specification for Application of Geosynthetics in Highway* (JTG/T D32).

2 When the height of retaining wall is more than 10 m, multi-level reinforced retaining wall shall be set. When the foundation of retaining wall is influenced by water flushing, gravity retaining wall shall be set below the flood level.

Fig. 5.1 Cross section of platform and bedding course

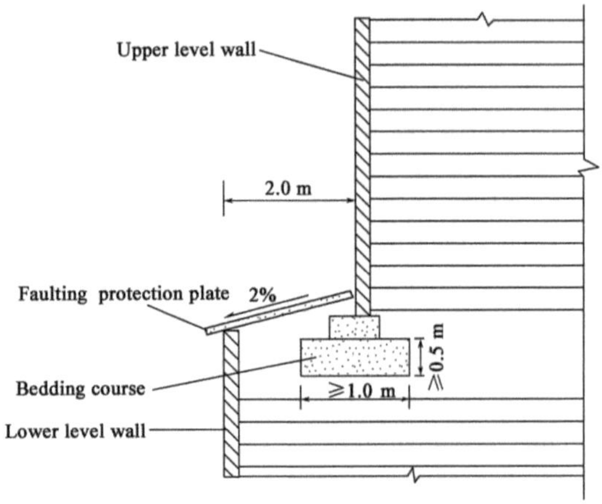

3 High density polyethylene geogrid and polyester welding geogrid should be used as geogrid in retaining wall.

4 The spacing between geogrid reinforcement layer, the length of steel, the slope ratio shall be determined by external and internal stability.

5 Filling and drainage design shall comply with the provision of Clause 5.4.10.

6 When the foundation is too weak and bearing capacity is insufficient, foundation soil shall be refilled and sand/ gravel cushion shall be set.

7 Reverse package geogrid shall use the same horizontal inflection length which shall be more than the value calculated from Eq. (5.4.11) and should be more than 2 m. The slope protection shall use anti-aging material.

$$L_0 = \frac{D\sigma_{hi}}{2(c + \gamma h_i \tan\delta)} \tag{5.4.11}$$

where:

L_0—horizontal inflection length (m);

D—spacing between steel (m);

σ_{hi}—lateral earth pressure stress (kPa);

c—cohesion between brace steel and filling (kPa);

δ—internal friction between brace steel and filling(°), (0.5–0.8)φ for sandy soil;

γ—density of filling (kN/m^3);

h_i—height from top of wall to the middle of level i plate (m).

5.4.12 The design of pile-plank retaining wall shall meet the following requirements:

1 he anchor pile of the pile-plank retaining wall must be anchored on a stable ground base, the cantilever length of the pile should not be larger than 15 m.

2 The pile structures may follow the requirements in Sect. 5.7 of the *Specifications*.

3 The lap length of the retaining plate lapped over the pile at each end shall not be less than the thickness of the plate. A convex platform for the lap shall be set at the

back of the circular pile. The width of the platform shall be 20–30 mm larger than the lap.

4 The concrete cover of the reinforcement on the front side of the retaining wall panel shall be larger than 35 mm and the concrete cover of the reinforcement on the back side of the retaining wall panel shall be larger than 50 mm. Stressed steel bars shall be placed continuously throughout the pile length. The stressed steel bars shall have a diameter of not less than 12 mm and a concrete cover of not less than 50 mm.

5 The arched retaining plate should not be only poured with concrete, but shall be also reinforced in both radial and circumferential directions with constructional steel bars. The bar diameter should not be more than 10 mm.

6 For the anchor pile with anchor rods, the deformation compatibility between the pile and the rods shall be ensured.

5.5 Slope Anchorage

5.5.1 In the process of slope anchorage design, the slope failure mode shall be identified and the degree and range of instability shall be determined based on the slope stability analysis data. The technical and economic demonstration of rationality and safety of the anchorage scheme shall be presented. The type of anchorage shall be determined based on the type of side slope rock soil mass, engineering characteristics, bearing capacity of anchor, material and length of the anchor as well as construction techniques, etc.

5.5.2 Prestressed anchor may be used for soil or rock slope reinforcement; the anchorage section shall be placed in a stable strata. The tensile anchor should be adopted for hard rock anchorage, while dispersed anchor should be adopted for soil and soft rock anchorage.

5.5.3 The stability evaluation of the prestressed anchor shall comply with the following requirements:

1 Anchorage slope stability evaluation shall comply with the requirements of Clause 3.7.4.

2 The methods of slope stability calculations used before and after the slope anchorage application shall be in correspondence.

3 In the calculation of anchorage slope stability, the anchorage force may be simplified as a concentrated force acting on the slope, or as a concentrated force acting on the sliding surface (see Fig. 5.2).

5.5.4 Design anchorage force shall be determined based on the slope instability force (sliding force) using Eq. (5.5.4).

$$P_{\mathrm{d}} = \frac{E}{\sin(\alpha + \beta)\tan\varphi + \cos(\alpha + \beta)} \tag{5.5.4}$$

where:

P_{d}—design anchorage force (kN);

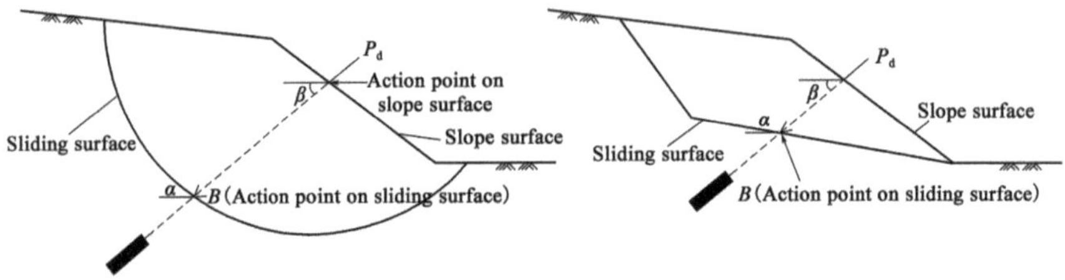

Fig. 5.2 Simplification of anchorage force

Table 5.5 Control stress for prestressing σ_{con}

Type of anchor	σ_{con}	
	Steel strand	Prestressed reinforced bar
Permanent	$\leq 0.50F_{ptk}$	$\leq 0.70F_{ptk}$
Temporary	$\leq 0.65F_{ptk}$	$\leq 0.80F_{ptk}$

E—sliding force (kN);

φ—internal friction angle of the sliding surface (°);

α—the dip angle of the sliding surface where the anchor and sliding surface intersect (°);

β—angle between the anchor and the horizontal plane (°).

5.5.5 The cross-section area of the anchor body shall be calculated according to Eq. (5.5.5). Control stress for prestressing σ_{con} shall comply with Table 5.5.

$$A = \frac{K_1 \cdot P_d}{F_{ptk}} \tag{5.5.5}$$

where:

A—cross section area of anchor body (m^2);

K—safety factor, selected according to Table 5.9;

F_{ptk}—tensile strength standard value of anchor body material (kPa).

5.5.6 Bond length of prestressed anchor shall meet the following requirements:

1 The bearing capacity of the anchor body is determined by three parts, including bond strength between grouting paste and anchor hole wall, bond strength between grouting paste and anchor and anchor strength. The smallest value shall be adopted for the design.

2 Prestressed anchors should adopt a bonded anchor body. Bond length between the stratum and grouting paste shall be calculated according to Eq. (5.5.6.1).

$$L_r = \frac{K_2 \cdot P_d}{\pi \cdot d \cdot f_{rb}} \tag{5.5.6.1}$$

where:

L_r—the bond length between stratum and grouting paste (m);

K_2—safety factor, selected according to Table 5.9;

d—boring diameter of anchorage section (m);

f_{rb}—bond strength between stratum and grouting paste (kPa). Bond strength shall be determined by a test. In conditions where tests are not available, Tables 5.6 and 5.7 may be referred to.

3 The bond length between the grouting paste and anchor body shall satisfy the requirements of Eq. (5.5.6.2).

Table 5.6 Characteristic values of the interface bond strength between rock mass and grouting paste

Type of rock mass	Saturated uniaxial compression strength R_c (MPa)	Bond strength f_{rb} (kPa)
Extremely soft rock	$R_c < 5$	150–250
Soft rock	$5 \le R_c < 15$	250–550
Comparatively soft rock	$15 \le R_c < 30$	550–800
Comparatively hard rock	$30 \le R_c < 60$	800–1200
Hard rock	$R_c \ge 60$	1200–2400

Note 1. Data in the table are applicable to the strength grade M30 of grouting
2. Data in the table are applicable only to the preliminary design and shall be checked through experiments when applied in construction
3. When structural planes of rock mass have developed, the lower limit values of the table is adopted

Table 5.7 Characteristic values of bond strength between soil mass and anchorage body

Type of soil mass	State of soil	Bond strength f_{rb} (kPa)
Cohesive soil	Stiff	60–80
	Stiff plastic	50–60
	Soft plastic	30–50
Sandy soil	Loose	90–160
	Slightly dense	160–220
	Medium dense	220–270
	Dense	270–350
Gravelly soil	Slightly dense	180–240
	Medium dense	240–300
	Dense	300–400

Note 1. The data in the table is applicable for grout strength grade M30
2. Data in the table applies only to the preliminary design and shall be checked through experiments when applied in construction

$$L_g = \frac{K_2 \cdot P_d}{n \cdot \pi \cdot d_g \cdot f_b} \tag{5.5.6.2}$$

where:

L_g—the bond length between stratum and grouting paste (m);

d_g—the diameter of anchor material (m);

f_b—the bond strength between grouting paste and anchor (kPa);bond strength shall be determined through experiments; when experimental conditions do not exist, reference to Table 5.8 may be used;

n—the number of anchor body.

4 The total anchor length consists of the lengths of anchorage section, free anchorage section and exposed section. The determination of the length of each part shall meet the following requirements:

(1) When determining the anchor bond length, length L_r and L_g shall be calculated separately and the bigger value of the two shall be adopted as the actual bond length, which shall be less than 3 m nor should it be greater than 10 m.
(2) The length of the free anchorage section is controlled by a stable formation interface. In design, the length of the free anchorage section inserted into the sliding surface or potential sliding surface shall not be less than 1 m, and the length of the free anchorage section shall not be less than 5 m.

5 During anchor design, the determination of the safety factor shall meet the requirements in Table 5.9.

5.5.7 The prestressed anchor rod structure shall meet the following requirements:

1 The prestressed anchor rod consists of an anchorage section, free anchorage length and anchor head. The anchor head consists of a saddle, steel backing plate and anchorage device.

2 The prestressed rebar in the anchorage section shall be provided with isolation frames every 1.5–2.0 m. The thickness of the protective coating of prestressed reinforcement shall not be less than 20 mm. For a temporary anchor, the thickness shall not be less than 10 mm.

Table 5.8 Bond strength design value between grouting paste and steel bars or strand f_b (MPa)

Type of anchor	Strength grade of cement slurry or cement mortar	
	M30	M35
Between cement mortar and deformed rebar	2.40	2.70
Between cement mortar and steel strand, high-strength steel wires	2.95	3.40

Note 1 When two steel bars are spot-welded into a bundle, the bond strength shall be multiplied by a reduction coefficient of 0.85

2. When three steel bars are spot-welded into a bundle, the bond strength shall be multiplied by a reduction coefficient of 0.7

Table 5.9 Design safety factor of anchorage body of prestressed anchor

Safety factor	Highway classification	Safety factor	
		Anchor service life ≤ 2 years (temporary anchor)	Anchor service life > 2 years (permanent anchor)
K_1	Motorway, Class-1 highway	1.8	2.0
	Under Class-2 highway	1.6	1.8
K_2	Motorway, Class-1 highway	1.8–2.0	2.0–2.2
	Under Class-2 highway	1.5–1.8	1.7–2.0

Note 1. If there is key protection object nearby, the safety factor may be adopted according to motorway
2. For soil mass or completely weathered rock mass, larger values shall be adopted for K_2

3 Anchor material may adopt high strength and low relaxation steal strand, prestressed reinforced bar according to the property of anchoring engineering, section of anchor and scale of engineering.

5.5.8 Anti-corrosion method shall be selected according to the service life and test of the corrosive environment of anchor rod. Anti-corrosion shall meet the following requirements:

1 Double-layer anti-corrosion shall be adopted for the permanent anchor. Simple anti-corrosion methods may be adopted for the permanent and temporary anchors which are not in the corrosive environment. Anti-corrosion I, II method shall comply with Table 5.10.

2 It shall be taken into account that anti-corrosion features will not be damaged during the anchor construction period and usage period when the right anti-corrosion methods are chosen for the anchorage section, free anchorage section and anchor head.

3 After completing the tension operation of anchor bolt, anti-corrosion protection shall be applied to relevant parts in anchor head. Steel shield should be set to anchor head which is prestressed and be filled with anti-corrosive grease. For anchor bolt which has been prestressed, the anchor head, bearing plate and the end geotextile may be protected with concrete and the thickness shall not be less than 50 mm.

5.5.9 Fully bonded anchor rod without prestress shall meet the following requirements:

1 HRB400 steel bars should be adopted as rod material and the diameter of the anchor bar should be 16 mm to 32 mm.

2 Boring diameter should not be less than 42 mm nor should it be larger than 100 mm.

3 The thickness of the protective coating of anchor bar shall not be less than 8 mm when cement mortar is adopted and shall not be less than 4 mm when resin is adopted.

Table 5.10 Anti-corrosion requirements for anchor I, II

Level of anti-corrosion	Type of anchor	Requirements for prestressed anchor and anchorage device		
		Anchor head	Free anchorage section	Anchorage section
I	Tension type separate-stretched	Using transition tube and anchor is protected by concrete or steel	Sheath injected grease or unbound steel strand or coated by tube unbound steel strand should be adopted	Bellows injected slurry
	Pressure type pressure-dispersive	Using transition tube and anchor can be protected by concrete or steel	Unbound steel strand	Unbound steel strand
II	Tension type separate-stretched	Using transition tube and anchor is protected by concrete or steel	Sheath injected grease or unbound steel strand	Slip casting

4 The construction measures of centering the rod shall be taken for anchor bars with length greater than 4 m or diameter larger than 32 mm.

5.5.10 The type of anchorage slope surface structure shall be comprehensively determined according to geological and hydrogeological conditions of the slope, properties of the rock and soil, slope height and construction method. The type of surface structure and application conditions shall be selected according to the requirements specified in Table 5.11.

5.5.11 The designs of slope structure shall conform to the following requirements:

1 A rectangular or T-shaped cross-section may be adopted for the ground beam and single-anchor pier; the thickness of the cross section shall not be less than 0.30 m. Frame beam may be designed to be rectangle or diamond. The size of rectangular

Table 5.11 Commonly used types of slope structure and application conditions

Type of structure	Applicable conditions	Notes
Lattice (frame) beam	Severe weathering, rich groundwater, soft rock, soil side slope	In rainy regions, the beam should be made to intercepting the ditch
Ground beam	Alternation of soft and hard rock, soil side slope	–
Single anchor pier	Hard rock, rock blocks or rock mass with good intactness	–

beam should not be less than 3 m × 3 m. The size of diamond beam should not be less than 5 m × 3 m.

2 The frame beam design should be carried out in sections, and the bending moment and shear force shall be calculated as a simply-supported beam or continuous beam. The loading design of ground beam and single anchor pier shall use the distance between two ground beams or the midpoint between two single anchor piers. The bending moment and shear force of ground beam shall be calculated based on number of anchors on the beam, and calculated as simply supported beam or continuous beam. The beam structure design shall follow the current *Code for Design of Concrete Structures* (GB 50,010) with the beam structure importance coefficient of 1.0 and permanent load sub-factor of 1.35.

3 Design of single anchor pier shall be in accordance with anchor force and shall meet the loading requirements of rock mass. The anchor pier shall be configured with the appropriate amount of structural steel.

4 The depth of beam bottom embedded into the rock slope should not be less than 0.20 m.

5.5.12 Anchor test and monitoring design shall meet the following requirements:

1 The anchorage test consists of a basic test and an acceptance test. In the early period of anchorage work construction, an anchorage test of the prestressed anchor shall be conducted. The number of basic tests for anchor bars is limited to 3% of working anchor bars and is not less than 3. After construction, an acceptance test shall be conducted. The number of acceptance tests for anchor bars can be limited to 5% of working anchor bars and is not less than 3. The number may be proportionately increased if there are special requirements. The content and requirements of the anchorage test shall comply with the requirements of the *Specifications for Bolt-shotcrete Support* (GB 50,086).

2 In-situ monitoring is conducted during the construction and permanent operating period. The number of monitored bars shall be 10% of working anchor bars. The project and method of monitoring may be determined according to Appendix F.3. The monitoring sections during construction shall be set at key parts of the anchorage zone. Operating period shall start at least one year after the completion of highway.

5.6 Soil Nailing

5.6.1 The design of soil nailing shall meet the following requirements:

1 Soil nailing may be applicable to the temporary support and permanent support of the cut slope of hard plastic or hard clay, cemented or weakly cemented silty clay, sandy soil, gravel, soft rock and weathered rock strata, etc. Permanent soil nailing should not be set for a cut slope with developed groundwater or in cohesive clay, soft plastic and fluid plastic clay.

2 A perfect drainage system shall be set on the ground surface and within the supporting structure to divert surface runoff and groundwater for the design of permanent soil nailing.

3 The design of permanent soil nailing shall be conducive to plant growth and in harmony with the surrounding environment.

4 Soil nailing shall adopt dynamic design and information construction. The horizontal displacement of the soil nailing slope shall not be larger than 0.3% of the slope height. If necessary, reinforcement measures shall be adopted.

5.6.2 Design of soil nailing shall meet the following requirements:

1 Soil nail support should be used for soil slope protection that is not higher than 10 m and for rock slope protection that is not higher than 18 m. In the case of high slope, multi-level soil nailing should be set. Platforms with width that should not be smaller than 2.0 m shall be set between the upper and lower levels of the multi-level slope with the height of each level should not be greater than 10 m.

2 The length of the soil nail consists of non-anchorage length and effective anchorage length. Non-anchorage length shall be determined according to the actual distance between the wall surface and potential failure surface of soil nails. Effective anchorage length is determined according to the calculation of the internal stability of soil nailing. The length of soil nail should be 0.5–1.2 times of the slop height. The space between soil nails should be 0.75 m to 3 m and the horizontal angle should range from 5° to 25°.

3 HRB400 steel bars with diameter 18 mm to 32 mm should be adopted for soil nail materials. The boring diameter should be 70 mm to 100 mm. A positioning frame shall be set for soil nail bars.

4 When the soil nails are used for corrosive environment, reinforced epoxy coating may be adopted. The thickness of protection shall not be less than 30 mm. Polyethylene plastic corrugated casing, polypropylene plastic corrugated casing or an epoxy-coated steel bar may be selected based on actual conditions at all the full length.

5 Cement paste or cement mortar should be adopted for the hole grouting material. The strength of the grouting material shall not be less than 20 MPa. The borehole method shall be adopted for grouting and the grouting pressure should be 0.4–1.0 MPa.

6 The thickness of shotcrete surface and size of steel concrete frame shall be calculated according to loading. The thickness of shotcrete surface should not be less than 80 mm.

7 Soil nails shall be effectively connected with the surface course. The connection method shall be determined according to loading and the importance. If necessary, it may be confirmed through loading tests.

5.6.3 The design of soil nailing shall meet the following requirements:

1 Soil nailing support structure calculation includes internal and external overall stability checking, support surface course, and the connection between the surface course and soil nail.

2 The checking calculation of external overall stability of soil nailing support may be done according to the method in Sect. 3.7; for the soil nail retaining structure, the checking calculation of sliding, overturning and bearing capacity of the soil nail reinforced soil mass may be conducted according to the requirements of Clause 5.4.2.

3 The arc method may be used to check the internal stability of the soil nail with the assumption that all soil nails in the failure surface are only subject to tension and have reached the maximum design tensile strength. The safety factor for checking the internal overall stability may be selected from a range of 1.25–1.30. In the case of seismic effect, the safety factor may be reduced by 0.1.

4 Bending strength and punching strength may be calculated based on continuous plate supported on soil nail in concrete slope.

5 When the slope protection is concrete frame beam or slab beam, internal force analysis and calculation shall be based on continuous beam system or slab beam system.

6 Local concrete bearing capacity calculations shall be conducted in the joint of soil nail and concrete slope.

5.6.4 In-situ test of soil nail shall meet the following requirements:

1 Before construction, a basic soil nailing pull-out resistance test shall be carried out for soil nailing works to determine its ultimate load and verify the limit bond strength of soil nailing interface. The number of tested soil nails shall not be less than 3. The construction method shall be the same with working soil nailing.

2 Soil nailing acceptance tests shall be determined in a random way. The number of test shall be 1% of the total working soil nails, and shall not be less than 3.

3 Creep tests shall be conducted for the permanent soil nailing in clay with plasticity index$I_p \geq 20$ and the liquid limit $w_L \geq 50\%$; for each test group there shall be at least 3 soil nail samples.

5.6.5 Monitoring of soil nailing test shall meet the following requirements:

1 In-situ monitoring during construction and permanent operating period shall be conducted for the working conditions and retaining effect according to the importance and actual condition of slope works. Monitoring items may be selected according to Appendix F. Monitoring locations shall be set on key locations of soil nailing areas.

2 Monitoring period during operation shall be determined by road grade and geological complexity. For cutting slope in highway, it shall not be less than one year.

5.7 Slide Resistance Pile

5.7.1 The design of slide resistance pile shall meet the following requirements:

1 Slide-resistance pile may be used to stabilize slope, landslide, unstable mountain and other special substructure.

2 Anti-slide piles should be set at sections with a relatively thin slide, small thrust force, and higher foundation strength of the anchorage section. It shall take comprehensive consideration to determine the layout of piles, pile spacing, pile length and section sizes to achieve economy, rationality and harmony with surrounding landscape.

3 Dynamic design and information construction shall be adopted for slide-resistance. Geological investigation and slide resistance shall be verified and

perfected according to the monitoring information of geological conditions and changes of slope in the process of pile foundation excavation.

5.7.2 When the pile cantilever section of soil landslide is overhead, soil retaining plate may be set between piles. If necessary, steel concrete shall be used to connect beams between slide resistance piles.

5.7.3 Slide resistance pile may be combined with prestressed anchor cable to make anti-slide retaining structure. The anchorage segment shall be set in stable rocks. The deformation of anti-slide pile in prestressed anchor cable shall coordinate with prestressed anchor cable. Anchor cable shall not work in shear state.

5.7.4 The design of slide-resistance structure shall meet the following requirements:

1 A rectangular section should be selected as the section shape of anti-slide pile. The cross-sectional dimension of pile shall be determined according to factors such as landslide thrust force, pile spacing, allowable lateral strength of anchorage section foundation, etc.

2 Pile body is made of concrete and the steel should be HRB400.

3 Locking wellhead shall be set at the wellhead of the anti-slide pile. Wall protection should be set when the pile well is located in soil strata and rock strata which are weathered and cracked.

4 The diameter of the longitudinal tensile reinforcement in the anti-slide pile shall not be less than 16 mm. Bar clear spacing should not be less than 120 mm. In difficult circumstances, the clear spacing may be reduced but shall not be less than 80 mm. When bundled bars are used, there shall not be more than three bars in a bundle. No more than three rows of reinforcement steel may be arranged if it is impossible to use single-rowed steel. The thickness of concrete cover for stressed reinforcement shall not be less than 70 mm.

5 The calculation of longitudinal reinforced cutoff point shall be according to the current *Code for Design of Highway Reinforced Concrete and Prestressed Concrete Bridges and Culverts* (JTG D62).

6 Oblique steel should be set inside anti-slide pile. The shear strength of inclined section may be met through adjusting diameter of stirrup, spacing and pile section size.

7 Enclosed type stirrups should be adopted. The diameter of stirrups should not be less than 14 mm and stirrup spacing shall not be larger than 0.40 m.

8 For the two lateral sides and compressive side of the anti-slide pile, longitudinal structural reinforcement shall be properly allocated with structural reinforcement spacing of 0.3 m, and its diameter should not less than 12 mm. At both lateral sides of the compressive side, framing bars with the diameter which should not be less than 16 mm shall be provided. For the long pile shaft, the diameters of both the longitudinal structural bars and framing bars shall be increased.

9 The bearing capacity of anchor head of prestressed anchor cable and other parts shall correspond with the largest tensile stress and tension technology. The distance between pile top and anchor hole shall not less than 0.5 m. The anchor cable structure shall follow Sect. 5.5. The force in concrete cushion block shall be uniform and in correspondence with cushion structure. The concrete cushion block

shall be well combined with pile structure. The local compression pressure shall be calculated according to *Code for Design of Highway Reinforced Concrete and Prestressed Concrete Bridges and Culverts* (JTG D62).

5.7.5 The design of anti-slide pile structure shall meet the following requirements:

1 The external forces acting on the anti-slide piles include landslide thrust, sliding body resistance in front of the pile and stratum resistance in the anchorage section. The pile shaft surface friction resistance, cohesion, pile gravity and pile bottom reaction force may be neglected. Landslide thrust shall be determined in accordance with the transfer coefficient method specified in Sect. 7.2.

2 The resistance force in front of the anti-sliding pile may be determined by the landslide thrust when the sliding body in front of the pile is in a limit equilibrium state, or by passive earth pressure in front of the pile. A smaller value is adopted.

3 The landslide thrust distribution graph on anti-slide pile shall be determined between rectangular or trapezoidal based on characteristics and thickness etc. of the slope. If the sliding body contains loose soil, triangle distribution may be adopted.

4 The pile bottom support should be assumed to be a free end or either a free end or may be assumed to be a hinged end if the pile bottom end is deeply embedded into the bedrock.

5 Anchorage length of the anti-slide pile shall meet the requirement that maximum compressive stress along the pile side is not larger than the lateral allowable bearing capacity of foundation.

6 The internal force of the pile above the sliding surface shall be calculated based on the landslide thrust and sliding mass resistance in front of the pile. Shaft displacement and internal forces of the pile below the sliding surface shall be calculated with the bending moment and shear force acting on the sliding surface. The foundation coefficient may be comprehensively determined according to the test data, regional experience and engineering analogies.

7 Anti-slide pile concrete design shall be in accordance with the current national standard *Code for Design of Highway Reinforced Concrete and Prestressed Concrete Bridges and Culverts* (JTG D62). The structure importance coefficient shall be 1.0 and the permanent load partial coefficient shall be 1.35. Anti-slide pile shaft is designed as a bending member. It may be not required to check deformation, cracking and deflection of the pile when there are no special requirements.

8 Prestressed anchor cable may be taken as elastic pile. Single-point anchor pile may be set as static system or statically indeterminate system. Internal force and displacement may be calculated through elastic deformation of pile anchor and foundation under loading. The design of prestressed anchor cable shall be in accordance with Sect. 5.5.

5.7.6 Monitor of slide resistance test shall meet the following requirements:

1 In-situ monitoring shall be conducted during the construction and permanent operating period. The content of monitoring may be determined based on Appendix F.3.5 typical monitoring site of stress and strain should be picked along the pile body.

2 The tests and monitoring of prestressed anchor cable shall comply with Clause 5.5.12.

3 Monitoring period during operation shall be determined by road grade and geological complexity. For high slope and landslides in highway, it shall not be less than one year.

Chapter 6
Substructure Widening and Reconstruction

6.1 General Requirements

6.1.1 Prior to design, investigations, surveys and tests shall be executed on the existing road substructure and the areas involving proposed substructure widening in terms of the performance of fill materials, moisture content, density, compaction, strength, and stability. The impacts of the widening or rebuilding substructure on the settlement and slop stability of the existing substructure shall be analyzed and evaluated.

6.1.2 For the substructure widening and reconstruction, a plan for utilizing and widening of the existing substructure shall be developed by comprehensive option comparisons in accordance with topographic and geological characteristics along the route, traffic composition before and after the widening, so that suitable and feasible measures can be adopted and strength and stability of the substructure after the widening can be secured. Rational engineering measures to ensure the strength and stability of the widening and reconstructing substructure shall be utilized.

6.1.3 For the substructure widening and reconstruction, the processing of existing substructure shall be comprehensively determined by taking the strength of existing substructure into account and according to the resilient modulus, rate of water content, density of the existing substructure.

6.1.4 For the substructure widening and reconstruction, attention shall be paid to the integrated design of substructure and pavement. The new substructure shall maintain a good connection with the existing substructure and some necessary engineering measures shall be taken to reduce the differential settlement between the new and existing substructure to prevent longitudinal cracks.

W. Wu, *Specifications for Design of Highway Substructure*, JTG Standards and Specifications, https://doi.org/10.1007/978-981-96-2313-6_6

6.2 Investigation and Evaluation of the Existing Substructure

6.2.1 In an existing substructure investigation, data collection shall be performed in an integrated way together with site investigations and geotechnical tests. Prior to the substructure widening and reconstruction design, the data of exploration and design, drawings and maintenances of the existing ground and substructure shall be collected. The settlement data of the existing road also shall be collected for the road built in soft soil areas.

6.2.2 The site investigations shall be conducted by means of road survey, nondestructive examinations, exploration tests and other relevant technical methods to identify the performance of the existing substructure, including drainage, protection works, retaining structures and so on. The site investigations shall meet the following requirements:

1 According to the existing data and the results of highway condition investigations, divide into sections and measure and evaluate for every section of the existing substructure.

2 Select typical sections to measure the geometry and dynamic deflection and perform bearing-plate tests to determine the resilient modulus of substructure. All the tests shall meet the related regulations of the current *Field Test Methods of Subgrade and Pavement for Highway Engineering* (JTG E60).

3 Typical crosses and sections with diseases shall be selected to explore and test to the pavement structures, soils of substructure and ground. The drilling depth and sampling tests shall meet the related regulations of the current *Code for Highway Engineering Geological Investigation* (JTG C20).

4 The foundation forms, geological conditions and service conditions of ground of the supporting structure of the existing substructure shall be investigated. The exploration tests for the ground of the supporting structure shall be performed if necessary.

6.2.3 The physical–mechanical properties tests of the roadbed soils of the existing embankment and cutting shall be performed to determine the rate of water content, degree of saturation, degree of compaction, average consistency, resilient modulus and CBR of the substructure soils etc.

6.2.4 The analysis and evaluation of the existing substructure shall include:

1 According to the investigations, measurements, tests and analysis of hydrological data, determine whether the elevation of the existing substructure meets the requirements for substructure design flood frequency specified in Clause 3.1.3.

2 Determine whether the fillings of the existing substructure meets the requirements of the minimum value of CBR and degree of compaction.

3 Determine the equilibrium moisture content in substructure, analyze and evaluate the rationality of the critical height of substructure.

4 Analyze and evaluate the slope stability and the effectiveness and improvement measures of various drainage facilities.

5 Analyze and evaluate the disease types, distribution scopes, scales and causes of the existing substructure and effects of facilities of disease treatment engineering for the existing substructure, and provide treatment measures of substructure diseases.

6.2.5 The analysis and evaluation of the existing substructure in soft soil area shall include:

1 Analyze and evaluate the degree of consolidation, consolidation coefficient, development law of compression deformation and increasing law of shear strength of the soft ground in various ground treatment sections under the existing substructure. Determine the degree of consolidation and residual settlement value (including both the primary consolidation and secondary consolidation) of the soft ground in various ground treatment sections under the existing substructure.

2 Analyze and evaluate the effect of the existing soft ground treatments and propose improvement measures.

3 Analyze and evaluate the stability and differential settlement between the widening and reconstructing substructure and existing substructure, the influence of the widening and reconstructing substructure on the settlement and stability of the existing substructure.

6.2.6 The analysis and evaluation of the existing substructure in expansive soil area shall include:

1 Determine the expansive properties and its depth and thickness, analyze and evaluate whether the substructure meets the related regulations in Sect. 7.9 of the *Specifications.*

2 Analyze and evaluate the stability and deformation forms of the existing substructure and the effectiveness of various treatment measures of expansive soils and improvement measures.

6.2.7 The analysis and evaluation of the existing substructure in karst area shall include:

1 Analyze the characteristic and scope of karst. Determine whether the save thickness of roof of karst cave and the distance from the karst cave to the substructure meet the related regulations in Sect. 7.6 of the *Specifications.*

2 Evaluate the stability and deformation forms of the existing substructure and the effectiveness of various treatment measures of karst and improvement measures.

3 Distinguish the influence of the widening and reconstructing substructure on the stability of the existing substructure.

6.3 Substructure Widening and Reconstruction of Class-2 and Lower-Class Highways

6.3.1 An appropriate type of cross sections for substructure widening and reconstruction shall be selected according to highway classification and technical standards and by taking local topography, geology, hydrology, earthwork requirements into account.

6.3.2 The levels of the substructure widening and reconstruction shall comply to the requirements specified in Clause 3.1.3 of the *Specifications*. Where the substructure height is restrained by terrain and cannot meet the requirements in Clause 3.3.2, the measures such as adding drainage cushions under the substructure or French drains underground shall be utilized.

6.3.3 For substructure widening, the foundation and ground treatment, the minimum strength and degree of compaction of substructure fill materials shall be in compliance with the technical requirements of corresponding classified highway after reconstruction. Where a Class-2 highway is to be reconstructed, the strength of the existing substructure may be upgraded as required.

6.3.4 Roadbed widening and reconstruction shall meet the following requirements:

1 The fillings used for widening and reconstruction of embankment should be the same as those of the existing embankment and meet the related requirements, or have a higher permeability than those of the existing embankment. When fine-grained soils are used for fillings, the drainage design between the new and old substructures shall be performed. If necessary, a leak ditch may be used to drain the water within the embankment.

2 For the embankment widening, benching shall be cut on the existing slope surface with a width no less than 1.0 m. When the widening width is less than 0.75 m, some engineering measures, such as filling the substructure with an extra width or scarifying the existing substructure, may be adopted.

3 The type and rate of side slopes of the widening embankment shall be determined in compliance to the requirements specified in Sect. 3.3.

6.3.5 For the cutting widening, the type and ratio of the slope may be determined according to the requirements specified in Sect. 3.4 or based on the stable slope of the existing cutting. For the stable sections of the existing cutting whose diseases have been treated for many years, the demolition projects shall be decreased and the existing slope should not be touched when reconstructing substructure.

6.3.6 For the reconstruction of a substructure with diseases, the treatment measures shall be used according to the disease types, characteristics, causes and hazard level of diseases combining with the factors of the local topography, hydrology, geology etc.

6.3.7 In places where the existing structures need to be altered because the substructure is raised or lowered or because the centerline is changed, the existing retaining structures should be retained if they are in good condition.

6.3.8 Existing buildings without significant damages and with a required strength and stability for reconstruction shall be fully utilized; buildings with partial damages or buildings that cannot meet the reconstruction requirements may either be strengthened to be utilized, reconstructed, or be demolished and rebuilt.

6.3.9 For the existing structures strengthened and utilized, the new and old concrete or masonry shall be closely linked to form an integrated body.

6.4 Widening and Reconstruction of Existing Substructure for Motorways or Class-1 Highways

6.4.1 A design of substructure widening and reconstruction shall meet the requirements specified in Sect. 6.3. An integrated design of ground treatment, substructure fillings, slope stability, drainage facilities shall be performed, which also need to coordinate with the designs of traffic engineering and pavement drainage system.

6.4.2 The degree of compaction of widened substructure shall meet the requirements of Clauses 3.2.3 and 3.3.4. For the design of the connection of the new and old substructures which shall meet the requirements specified in Clause 6.3.4, a geogrid shall be laid horizontally for an embankment with a height over 3 m to enhance the whole stability of the substructure and reduce the differential settlement.

6.4.3 The design of the substructure widening on a soft ground shall meet the related regulations in Sect. 7.7 and shall also meet the following requirements:

1 The differential settlement between the new and old substructures shall be controlled during the substructure widening and the post-construction increment of the crown cross slope of the existing and widening substructures shall not be greater than 0.5%.

2 For the selection and design of the ground treatment, several associated factors, such as the thickness and depth of the soft soil, the degree of consolidation and the residual settlement of the existing ground, the height of substructure and widening forms, shall be considered to control the settlement of the widening substructure and try to reduce the influence to the existing substructure.

3 For the shallow ground, the cushion and the shallow treatment methods may be used to reduce the settlement of the widening substructure.

4 For the deep and thick soft ground, the treatment measures such as composite ground or light embankment may be used, while the drained consolidation methods or dynamic compaction methods which have a serious influence to the existing substructure should not be used. For the sections with fish or water ponds, rivers, reservoir, etc. that require drainage and dredging, the water level shall not be lowered until seepage-proof and water-proof measures have been taken.

5 When the new substructure is separated from the existing substructure with a distance between them less than 20 m, the isolation measures or ground treatments for the new substructure may be used to reduce the influence to the settlement of the newly built substructure on the existing one.

6.4.4 For the existing substructure in sections with poor hydrological conditions, an additional drainage cushion and underground drainage ditch etc. shall be provided combining with the substructure and pavement widening and reconstruction design.

6.4.5 The traffic organization design shall be performed well for the substructure widening of an motorway or Class-1 highway. For the rock cutting, the smooth blasting or presplit blasting shall be used and some relative protection measures shall be taken.

6.4.6 The use of the existing substructure shall be combined with the use and the overlay design of the pavement. At the same time, the following treatment measures shall be used according to the causes of the substructure diseases and the influence to the widening structure:

1 When the resilient modulus of the existing substructure cannot meet the requirements of the newly-built substructure but the existing pavement with no damages may meet the requirements of the pavement design after widening by the overlay design, the existing should suitable to be utilized fully.

2 When the resilient modulus of the existing substructure cannot meet the requirements of the newly-built substructure and the existing pavement has a serious damage, the treatment measures, such as drainage facilities improving, supplementary rolling, filling replacing etc., may be used according to the analysis and evaluation of the rate of water content, degree of compaction and type of filling.

3 When the existing substructure cannot be excavated with the limited conditions, the treatment measures, such as the cement gravel pile, cement fly-ash gravel pile, and plaster casting method etc., may be used.

6.4.7 When a Class-2 or below highway is widened and reconstructed to an motorway or Class-1 highway, soils of the existing substructure shall be improved or the existing substructure and pavement shall be excavated and refilled when the strength and the degree of compaction of the existing substructure cannot meet the related requirements and the plan of reinforcing pavement is not feasible as a whole after demonstration.

Chapter 7
Special Substructure

7.1 General Requirements

7.1.1 Special substructure includes special soil (rock) substructure, poor geological substructure and substructure in special conditions. As the route passes through the special section, a comprehensive geological survey shall be performed to identify the nature of the special geological body, origin types, sizes, stability status and development trends. The physical and mechanical parameters of the special substructure design shall be comprehensively analyzed and determined by the in-situ test data combined with laboratory test data.

7.1.2 A geological investigation shall be conducted along the road alignment. The alignment shall avoid the areas with massive, complex and difficult geological conditions and special soils (rock) as well as highly backfilled and deeply excavated embankments.

7.1.3 In the special substructure design, the effects of geological, water and environmental factors in long-term performance of substructure shall be considered. In a substructure disease treatment, the principle of case-base, combining prevention with control and seeking to cure shall be followed. By comprehensive technical and economic comparison, a reasonable treatment scheme and effective engineering measures are implemented with local conditions. If treatment is applied in phases, the substructure safety shall not be compromised in the change process of various factors.

7.1.4 Dynamic design should be applied in motorway and Class-1 highway design.

7.2 Substructure in Landslide Areas

7.2.1 Substructure design in landslide section should be according to the following principles:

W. Wu, *Specifications for Design of Highway Substructure*, JTG Standards and Specifications, https://doi.org/10.1007/978-981-96-2313-6_7

1 For substructure design in landslide sections, the properties of landslide, topography, geomorphology, hydrogeology and engineering conditions near the landslide mass as well as landslide genetic types, sizes and characteristics of the landslide shall be identified. The stability condition and development trend of the landslide and its extent of damage to highway engineering shall be analyzed and evaluated. Effective measures shall be taken in a timely manner to ensure the safety of substructure construction and operation.

2 For landslide in large scale, with complex features, slow deformation and slide features that are difficult to detour in the short term, the program of overall planning and phased treatment may be adopted.

3 For landslide control, integrated management measures such as drainage, load reduction, counter pressure and support engineering shall be taken according to the types, sizes and stability of the landslide, combined with engineering geological conditions of landslides, highway importance, construction conditions and other requirements.

4 For a high side slope or slope with special soil and an adverse structural plane, necessary prevention measures shall be taken to avoid engineering landslide.

7.2.2 The engineering geological analogy method and mechanical calculation shall be adopted to comprehensively evaluate the landslide stability. The design shall meet the following requirements:

1 The following three working conditions shall be considered in the calculation of landslide stability:

(1) Normal case: slopes with natural conditions;
(2) Special case I: slopes with heavy rain or continuous rain conditions;
(3) Special case II: slopes with different loads for example earthquake.

2 The factor of safety of slope stability shall not less than the values in Table 7.1.The slope stability calculation and the factor of safety for Special case II shall follow the current *Highway Engineering Seismic Design Code* (JTG B02).

3 The following loads shall be included in the landslides treatment design: permanent load including the gravity of sliding blocks, surcharge above the sliding blocks, the hydrostatic and hydrodynamic water pressure caused by groundwater, vehicle loading; temporary load including earthquake, and construction on the sliding blocks.

4 The shear strength of the slide surface soil should be determined according to laboratory test data of the slip surface rock and soil, the limiting equilibrium inverse calculation value and engineering geology analogical experience data, combined with the worst condition of landslide. If necessary, it shall be determined by field test data.

5 The transfer coefficient method may be used to calculate the landslide residual sliding force according to Eq. (7.2.2.1).The stick force system is shown in Fig. 7.1. When $T_i < 0$, T_i shall be 0. When the residual sliding force of the last stick is greater than 0, the landslide is stable. When the residual sliding force of the last stick is less than 0, the landslide is not stable.

$$T_i = F_s W_i \sin\alpha_i + \psi_i T_{i-1} - W_i \cos\alpha_i \tan\varphi_i - c_i L_i \tag{7.2.2.1}$$

Table 7.1 Factor of safety of slope stability

Highway class	Factor of safety	
	Normal	Special case I
Motorway and class-1 highway	1.20–1.30	1.10–1.20
Class-2 highway	1.15–1.20	1.10–1.15
Classes-3 and 4 highway	1.10–1.15	1.05–1.10

Note 1. Slopes with complex and server conditions, the factor of safety may use higher value; when the geological condition is simple, the factor of safety may use smaller value
2. Landslide areas including important structures (bridges, tunnels, electricity transfer towers, and buried pipes), villages and schools may use higher value of factor of safety
3. In water reservoir areas, higher factor of safety of landslide may be used for the water reservoirs with frequent water level change and rapid drawdown
4. For temporary or emergency constructions, a factor of safety of 1.05 may be used for the design of treatment of landslides

$$\Psi_i = \cos(\alpha_{i-1} - \alpha_i) - \sin(\alpha_{i-1} - \alpha_i)\tan\varphi_i \qquad (7.2.2.2)$$

where:

T_i, $T_{i\text{-}1}$—the residual slide force of the ith and (i-1)th slide stick, respectively (kN/m);

F_s—the stability factor;

W_i—the weight of the ith slide slice (kN/m);

α_i,$\alpha_{i\text{-}1}$—the slip surface inclination angle that corresponds to the ith and $(i\text{-}1)^{th}$ slide slice, respectively (°);

Ψ_i— the transfer coefficient;

φ_i—the slip surface internal friction angle of the ith slide stick(°);

c_i—the slip surface cohesion of the ith slide slice (kN/m);

L_i—the slip surface length of the ith slide slice (m).

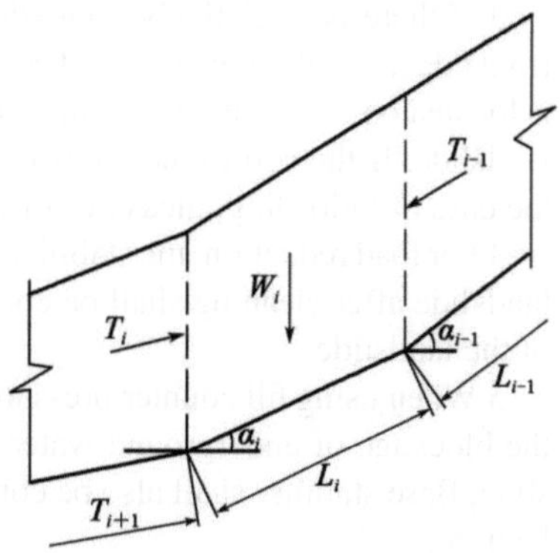

Fig. 7.1 Residual slide force system diagram

7.2.3 The design of landslide treatment should be based on the suitability of treatment methods and comparison of different treatments considering the economy which also meet the local requirements.

7.2.4 For the design of drainage engineering, the drainage program shall be established along with engineering geology, underground water and rainfall conditions according to the overall control program of landslide. The following requirements shall be met:

1 For ground surface drainage engineering, an annular intercepting ditch shall be set in the stable stratum of the landslide back-edge. In a large landslide area, a dendritic drainage ditch system should be set in the range of landslide mass. For places where drainage ditches pass cracks, crack control measures shall be taken. Slopes with significant cracks and deformation shall have cracks filled in time with cohesive soil or cement.

2 Drainage programs such as sewers, oblique drainage holes or tunnels shall be chosen for underground drainage engineering according to the conditions of slip surface, hydrogeological conditions of the slope catchment basin and dynamic characteristics of underground water.

3 The plane layout of the intercepting sewer shall be perpendicular to the direction of underground water flow and be built on stable soil 5 m away from the landslide area. A filter layer shall be set on the upstream face of the sewer and seepage insulation layer on a downstream surface.

4 Design of sewer, blind ditch, oblique drainage holes and tunnels shall follow the requirements in Chap. 4.

7.2.5 Load reduction and counter pressure measure shall meet the following requirements:

1 For a thrust load-caused landslide or scattered conversion landslide, the measures of reducing load at the back-edge of the landslide and counter pressure at the front-edge should be adopted.

2 When the landslide bed has the shape of an upper part steep slope and lower part flat slope, the stratum of the back-edge and both sides of the landslide are relatively stable and the landslide would not slide towards the back-edge nor both sides because of weight reduction by excavation, the measures of weight reduction should be taken.

3 If there is a relatively long sliding resistance section in front of the landslide front edge, weight reduction and counter pressure should be used. If the substructure is located on the front edge of the landslide, embankment should be used to pass the landslide. If the soil in the landslide mass or the slip zone has expansion cracks in the case of unloading, measures of weight reduction shall not be taken.

4 For load reduction, the stability of mountains in the back and on both sides of the landslide after cleaning shall be considered to prevent a new slide at the back-edge of the landslide.

5 When using fill counter pressure measures, attention shall be paid to preventing the blockage of underground water seepage channels in the front-edge of the landslide. Base stability shall also be considered and if necessary ground treatment shall be used.

7.2.6 Slide-resistance supporting and retaining structures shall meet the following requirements:

1 Slide-resistance retaining wall should be set at the front-edge of the landslide. If necessary, it may combine with measures such as drainage, weight reduction, anchorage and so forth.

2 Slide-resistance retaining wall shall be designed according to the larger value between residual slide force and coulomb earth pressure. The wall height and buried foundation depth shall be determined in order to prevent slip mass sliding over the wall top or under the foundation.

3 Design of slide-resistance retaining wall shall be in accordance with relevant requirements in Sect. 5.4 of the *Specifications*.

4 When the foundation of the slide-resistance retaining wall is very deep and soil stability is poor, temporary retaining measures shall be adopted. The construction must be carried out by sections to ensure the stability of landslide and construction safety during construction.

7.2.7 Slide-resistance pile should meet the following requirements:

1 The position selection of slide-resistance pile shall follow the requirement of Item 2 in Clause 5.7.1.

2 The slide-resistance pile should be mainly arranged in a single row. If the landslide thrust force is relatively large, subsection sliding resistance may be carried out against the landslide. If the moment is too large, a prestressed anchor (cable) slide-resistance pile shall be used.

3 The length of the slide-resistance pile should be less than 35 m. For landslides with slip zone depth of more than 25 m, the feasibility of using the slide-resistance pile to prevent the landslide shall be fully demonstrated.

4 The structural design of the slide-resistance pile shall be in accordance with relevant requirements in Sect. 5.7.

7.2.8 Prestressed anchor shall meet the following requirements:

1 Anchoring part of the prestressed anchor (cable) shall be placed in the stable stratum behind the slip surface.

2 Pressurized structures of prestressed anchor (cable) shall be determined according to the rock properties and bearing capacity of landslide mass. Reinforced concrete frames or ground beams should be used and measures to prevent the topsoil from being eroded by rainwater and local soil collapse shall be adopted for the slope surface.

3 Design of the prestressed anchor (cable) shall be in accordance with the relevant requirements of Sect. 5.5 of the *Specifications*.

7.2.9 When a river causes the front-edge of landslide mass to erode, scour protection measures shall be adopted.

7.2.10 For landslide control of Motorway and Class-1 highway, landslide monitoring and dynamic design shall be carried out, the design shall meet the following requirements:

1 The focus of landslide monitoring, which includes construction safety monitor, control effectiveness monitor and operation monitor, shall be placed on construction safety monitor and control effectiveness monitor.

2 The landslide monitoring project may be selected according to Appendix F. Monitoring points shall be arranged, forming a complete profile, in locations where the stability of landslide mass is poor or engineering disturbance is great.

3 Control effectiveness monitoring shall be carried out with integrating with construction safety and operation monitoring. The duration of control effectiveness monitoring shall not be less than one year after the renovation project has been completed and the road operated.

7.3 Substructure in Rockfall and Talus Sections

7.3.1 The design of the substructure in rockfall shall consider the following requirements:

1 For substructure design in sections of rock fall and talus, data that involves terrain, geomorphy, geology, hydrology and meteorology shall be investigated. The type, scope and causes of rock fall and talus, and extent of damage on the highway shall be ascertained to predict the occurrence and development of rock fall and talus after road completion and evaluate the stability and a comprehensive control measures shall also be considered.

2 In the design of substructure, high fill and deep cut shall be avoided and the route shall be away from the accumulation zone of rock fall materials. Sections of medium and small rock fall shall be comprehensively treated using engineering measures such as shading, interception, removal and reinforcement.

7.3.2 When the side slope or natural slope is relatively flat and the rock surface is easily weathered into small pieces of rock with sporadically falling, slope protection should be carried out to prevent these developments and sporadically falling of rock.

7.3.3 When there is not a large number of falling rocks on the hillside or side slope and fragmentation of rock is not serious, the fragmental rocks may be removed and the side slope may be flattened. The measures also include supporting structures and grouting anchor system with hanging net. Blocking structures such as a flexible protection system or a retaining wall and rockfall cistern may also be adopted. The retaining wall should be used in conjunction with the rockfall cistern and the position may be reasonably arranged according to topography. A buffer layer shall be set at the back of the retaining wall. The design of the retaining wall shall be in accordance with the highway retaining wall design and in consideration of the landslide impact load effect on the pressure the at wall back.

7.3.4 A substructure with dangerous rocks, the rocks shall be removed or other measures including supporting structures and prestressed anchors shall be used. For high side slope with weathered zones and fissure developing zones, the rock removing method should not be adopted.

7.3.5 When the rockfall mass is large, occurs frequently, and is close to the route but interception structures are difficult to set up, then blocking structures such as open-cut tunnels and shed tunnels may be used. Blocking structures shall have sufficient lengths and a buffer layer on the roof. The effect of stacked stone load and impact load shall be considered.

7.3.6 Design structures to prevent rockfall including type, dimension, number and locations shall be determined based on several factors of falling rocks including size, number, distribution, impact force and distance to the highway.

7.4 Substructure in the Section Backfilled with Rocks

7.4.1 Design of the substructure in the section backfilled with rocks shall use the following principles:

1 The topography, geomorphology and geology conditions of the rock filled sections shall be investigated. The material, type, distribution and cause of the rock pile should be understood to analyze the occurrence, developing and influence to the stability of the highway.

2 Design of the rock backfilled section shall base on the distribution, thickness, material, slope shape below the rocks, properties of the soils, groundwater, and surface water conditions to evaluate the stability of the rocks and select the location of the alignment and the substructure shapes.

3 The alignment shall avoid the huge rock piles with big area, steep slope, sufficient supply and low stability. For medium and small rock piles, the substructure shall use low embankment or excavation when the alignment is difficult to avoid the rock piles. The reinforcement should also be adopted.

7.4.2 Design of the substructure in the section backfilled with rocks shall be evaluated based on the location and cross section of the substructure. The factor of safety of stability should meet the requirement in Sects. 3.6 and 3.7.

7.4.3 The excavation shall be avoided in the developing rock filled sections. The reinforcing measures including retaining wall, surfaced slope, rock prevented wall, rock fall trench, open cuts and closed cuts should be adopted.

7.4.4 The following measures should be used in the stable rock backfilled substructures:

1 The benched substructure should be used and gentle the slope or remove the rocks above the substructure when the highway is above the rock piles.

2 The earth structures for example retaining wall should be constructed aside the excavation when the highway is going through the rock piles.

3 Backfill embankment should be adopted when the highway is below the rock piles.

7.4.5 The appropriate measures of reinforcement and blocking shall be adopted when the highway is going through the active rock piles. The measures shall be determined based on the area, rock type and size of the rock piles.

7.4.6 Anti-slide retaining wall and anti-slide piles should be constructed when the stability of the substructure backfilled with rocks is insufficient.

7.5 Substructure in the Section of Debris Flow

7.5.1 Design of the substructure in the section of debris flow shall adopt the following principles:

1 For the design of substructure in the section of debris flow, the genetic type, scale, characteristics, activities, laws, trends and extent of damage of debris flow shall be ascertained.

2 The highway alignment shall avoid the big size debris flow, debris flow groups, the heavily-deposited trench and heavily-deposited riverbeds. Proper alignment location, cross section and reinforcement shall be adopted when it is difficult to avoid medium and small debris flow.

3. In controlling debris flow, various measures such as row guidance, interception and soil and water conservation shall be taken into account. Overall planning shall be made to take comprehensive control.

7.5.2 When the highway is over the debris flow trenches, the structure shall be constructed at the flow area and stable zones. The zones with slope change from steep to gentle and upward sharp turning shall also be avoided. The design shall meet the following requirements:

1 Bridge may be used in crossing debris flow gullies of the circulation area or the stable natural groove of the alluvial sector. In design, a reasonable span and bridge type shall be designed according to topography, geology, erosion condition of gully, channel width, border spread of debris flow, mud wave height, flow, development trends, and so forth.

2 Tunnel may be used in crossing large-scale debris flow gully or many debris flow gullies that may cause severe damage to route. The tunnel plan shall be selected after technical and economic comparisons with other plans. The tunnel hole itself shall be set in the stable stratum at the bottom of debris flow, and the entrance and exit shall be away from the possible harm of debris flow.

3 Culvert should not be adopted in the area of debris flow nor shall it be adopted in the active alluvial fans of debris flow. For Class-3 and 4 highways, culvert may be adopted if the scale of debris flow is not large, the content of solid matter in the flow is low, the flow is without large stones and there is a straight groove.

4 Ford may be used in Class-3 or 4 highways across debris flow with small slope surface. The substructure transverse section of the ford shall be fully enclosed, and may be used in combination with bridges and culverts. The cut-off wall shall be set at the substructure toe to prevent scouring.

7.5.3 Drainage guiding ditch, aqueduct and alluvial fans shall be constructed to prevent debris flow and protect ditch when the alignment is passing the debris flow deposited zones. The design shall meet the following requirements:

1 Drainage guiding ditch may be used in roads in sediment terrain. Outlets shall be joined with the main channel while the elevation of outlet shall be higher than the elevation of 20 years flood water level. The longitudinal slope of the drainage ditch should be determined according to the ground slope. The cross-section of the

drainage ditch shall be determined by flow calculation. The drainage ditch shall be protected.

2 Aqueduct may be used in debris flow with a discharge flow less than 30 m^3/s, and where the topographic conditions shall allow the longitudinal slope of the aqueduct to meet the design requirements and provide the required road clearances. There shall be an area at the downstream of the substructure for stopping silt. Aqueduct shall be straightly and smoothly joined with the original groove. The longitudinal slope of the aqueduct shall not be smaller than the longitudinal slope of the original groove. The outlet shall meet the needs of discharging debris flow. The design load of the aqueduct shall be calculated by the full debris flow in the aqueduct and impact force. An impact factor of 1.3 may be taken.

3 For a certain range of alluvial fans where the direction of debris flow needs to be controlled or the scope being affected needs to be restricted, a diversion dike may be set to prevent debris flow from directly washing on the embankment or congesting bridge. The height of the diversion dike shall be the sum of the deposition depth in design service life and the furrow depth of debris flow. At the place where debris flows can be blocked or in curve, rising height and curve height shall be added.

7.5.4 Blocking structures including blocking dam and grid dam may be constructed to block the rocks and minimize the impact of the debris flow and depositing behavior when the alignment is going through the debris flow area. The design shall meet the following requirements:

1 Blocking dam may be suitable for terrain conditions where diversion of sediment and stopping silt are not feasible at upper and middle streams or downstream of valleys and where the sand carried from upstream has to be controlled and where, along the channel, there is a big sand catchment area, with intensive rockfall and landslide. The location of blocking dam shall be comprehensively determined based on the purpose of the dam setting, combined with the valley terrain and geological conditions of the foundation. The maximum dam height should not be more than 5 m and a flat top should be adopted. If the shore slope at both ends has scour potential, a concave shape should be adopted.

2 Grid dam may be used in a small debris flow with smaller interception flow and less large stone content. The grid interval in the grid dam is set to intercept large stones and discharge fine particles. The discharge section shall meet the requirement of safely discharging flood to downstream. The width of the dam shall be the same as the width of the groove. The dam base shall be set on solid subsoil.

7.5.5 In the design of the substructure in debris flow areas, the slope should be reinforced using plants which shall include bushes, trees and herbage.

7.5.6 The construction safety and protection shall be monitored when the alignment could not avoid huge debris flow. The monitoring shall include the frequency, flow rate and material of debris flow and the relationship of the change of flow rate with the river flow and rainfall.

7.6 Substructure in Karst Area

7.6.1 The design of substructure in Karst area shall meet the following requirements:

1 For the design of substructure in Karst area, effective survey methods such as remote sensing, geophysical exploration and drilling shall be adopted to obtain the data such as the geomorphology of karst, development of karst, development law and surrounding rock properties of the karst cave, as well as the activity law of surface drainage and underground water. The reason of ground collapse shall be analyzed and the stability should be evaluated. A makeup investigation and evaluation should be necessary when the excavation reaches the top surface of the substructure.

2 The highway alignment shall avoid the huge and complex Karst developing areas. The substructure should be constructed at a relative lightly-developed and easily-treated areas.

3 For substructure in a karst section, the extent of damage of the karst on substructure shall be judged based on the stability of the substructure and the environment influence along with engineering practice and a reasonable treatment method shall be chosen.

4 Substructure shall not cut the groundwater and surface water in the Karst areas and cause blocking and stopping water or shortage of water in farmer land in Karst developing areas.

5 The foundation reinforced using grouting, the coring shall be used to evaluate the effectiveness of the grouting.

7.6.2 When the roof strata of a karst cave has not been cut by a joint crack or has been cut but possess a complete roof which is well bonded, the safety thickness of the cave roof may be determined according to the thickness-span ratio. When the ratio of the roof thickness and the length of karst cave, which is spanned by substructure, is larger than 0.8,the roof strata of karst cave may not need to be treated.

7.6.3 Safe distance from karst cave to substructure shall follow the requirements as follows:

1 When karst topography is located on both sides of the substructure, the effect of karst on the substructure shall be determined. The stability and influence of open karst topography on the substructure may be determined based on the natural side slope. For an underground karst cave, the influenced range may be calculated by the diffusion angle in collapse (see Fig. 7.2) according to Eq. (7.6.3.1).

$$L = H\cot\beta \tag{7.6.3.1}$$

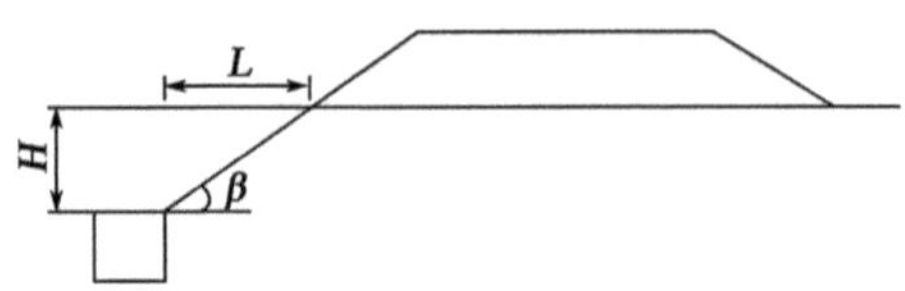

Fig. 7.2 The schematic diagram of safe distance of karst cave(L)

$$\beta = \frac{45^\circ + \frac{\varphi}{2}}{K} \tag{7.6.3.2}$$

where:

H—thickness of roof of karst cave (m);

L—safe distance of karst cave;

β—diffusion angle in collapse (°);

K—safety factor should be taken as 1.10–1.25; larger value shall be used for Motorway and Class-1 highway;

φ—internal friction angle of rock and soil (°).

2 If there is overlying soil on the roof rock, an oblique line could be drawn from the soil bottom to find an intersection with the ground. The slope toe shall be located outside of the intersection range and the distance shall not less than 5 m.

3 When the slope toe is located outside of the influenced area by collapse spread of karst cave, karst cave may not be dealt with.

7.6.4 Water in Karst areas, which can influence the stability of the embankment, shall be channeled and directed out. The measures shall meet the following requirements:

1 For karst springs and water-emitting holes at the upslope, a drainage ditch should be adopted to intercept and drain the water out of the substructure.

2 For karst springs and water-emitting holes at the substructure substrate, a catchment open channel or sewer should be set to discharge water out of the substructure.

3 If water outlets in Karst areas are partially blocked, the remaining open outlet shall meet the sufficient drainage requirement.

4 The surface water shall be collected and drained together.

7.6.5 When the substructure is located at a low karst area, a sufficient drainage system shall be installed to drain the surface water to the surrounding trenches or sinkholes without influence to the substructure. If the water could not be drained, the substructure shall be backfilled using material with good permeability including gravels or gravelly soil, and the substructure shall be 0.5 m higher than the water level.

7.6.6 The underlying karst cave in substructure area shall be evaluated based on size, depth, water condition, locations, construction condition of sinkholes. The following methods shall be adopted to reinforce sinkholes:

1 The karst holes and sinkholes shall not be blocked if the drainage is necessary. Structures including steel-reinforced concrete plates, bridges and culverts shall be constructed to span holes to protect groundwater in Karst areas.

2 The dry karst caves on the stable cutting side slope should be filled with dry-laid rubbles.

3 For exposure and shallowly-buried karst holes at the bottom of substructure, the measures including backfilling, spanning with steel-reinforced concrete plates and other supported structures may be adopted.

4 A karst cave with filling may be preferably reinforced by grouting or jet grouting. If the reinforcement cannot meet the design requirements, structures should be adopted to span it.

5 For shallowly-buried sinkholes, the reinforcement may include backfilling, impact compaction, dynamic compaction. The surface water need to be drained properly. For deeply-buried sinkhole, grouting and composite foundation should be used to reinforce sinkholes.

7.7 Substructure in the Area of Soft Soil

7.7.1 Design of the substructure in the soft soil area shall meet the following requirements:

1 Data involving topography, geomorphology, engineering geology, hydrology, meteorology, etc., along the line shall be surveyed according to relevant requirements in the *Specifications for Survey of Highway Engineering Geology* (JTG C20).A comprehensive exploration test and in-situ test shall be carried out by appropriate exploration methods and statistics, and an analysis shall be carried out to provide reliable physical and mechanical properties of the soft soil for design.

2 The factor of safety of stability of substructures over soft soils shall meet the requirements in Table 7.2. When the value of factor of safety is lower than the value in Table 7.2, ground improvement shall be adopted.

3 The post-construction settlement of substructure shall meet the requirement in Table 7.3. When the value of factor of safety is lower than the value in Table 7.3, ground improvement shall adopted.

7.7.2 Calculation of settlement of subsoil shall meet the following requirements:

1 The calculating depth of the settlement shall be the place with the ratio of the additional stress and effective overburden stress is not greater than 0.15.

2 The influence of traffic loading may be neglected for high embankments.

Table 7.2 Allowable factor of safety

Index	Consolidated stress solution		Modified total stress method		Simplified Bishop method and Janbu method
	With consolidation	Without consolidation	With consolidation	Without consolidation	
Direct shear	1.1	1.2	–	–	–
CPT and vane shear	–	–	1.2	1.3	–
Effective triaxial shear parameters	–	–	–	–	1.4

Note The factors of safety in the table minus 0.1 when the earthquake is considered

Table 7.3 Allowable post-construction settlement (m)

Highway class	Locations		
	Connection of abutment and embankment	Box culverts	General
Motorway and class-I highway	≤0.10 m	≤0.20 m	≤0.30 m
Class-II highway as main road	≤0.20 m	≤0.30 m	≤0.50 m

3 Primary consolidation settlement S_c shall be calculated by the layer wise summation method.

4 Total settlement S should be calculated by settlement coefficient m_s and primary consolidation settlement.

$$S = m_s S_c \tag{7.7.2.1}$$

where:

m_s—settlement coefficient, which is related to conditions of subsoil, load intensity, loading rate and other factors; the value ranges between 1.1–1.7; it shall be determined according to observation data of settlement in the field and also may be estimated with the Eq. (7.7.2.2);

$$m_s = 0.123\gamma^{0.7}(\theta H^{0.2} + vH) + Y \tag{7.7.2.2}$$

θ—the coefficient of ground treatment type; if the subsoil is treated with prefabricated strip drain plate, it is taken as 0.95–1.1; if the subsoil is treated with silt mixing pile, it is taken as 0.85;if the subsoil is treated with general preloading, it is taken as 0.90;

H—the height of substructure at center (m);

γ—the fill gravity (kN/m^3);

v—the correction factor of the filling soil rate; if the filling rate is 0.02–0.07 m/d, it is taken as 0.025;if using stage-loading and the filling rate less than 0.02 m/d, it is taken as 0.005; if using fast loading and the filling rate greater than 0.07 m/d, it is taken as 0.05;

Y—the correction factor of geological factor; when the undrained shear strength of soft soil is less than 25 kPa, soft soil thickness is greater than 5 m, and the hard shell layer thickness is less than 2.5 m, Y is taken as 0;otherwise, it may be taken as-0.1.

5 The total settlement may also be calculated by the summation of instant settlement S_d, primary consolidation settlement S_c and secondary consolidation settlement S_s, that is:

$$S = S_d + S_c + S_s \tag{7.7.2.3}$$

6 Considering the change process of primary consolidation with time, the subsoil settlement at any time is calculated according to the following equations:

$$S_t = (m_s - 1 + U_t)S_c \tag{7.7.2.4}$$

$$or S_t = S_d + S_c U_t + S_s \tag{7.7.2.5}$$

where:

U_t—the average of consolidation of subsoil which can be calculated by the theoretical solution of Terzaghi one-dimensional consolidation. For the foundation being treated with vertical drains such as sand drain, prefabricated strip drain plate and so forth, the consolidation is calculated by the solution of consolidation equation of the Terzaghi-Rendulic consolidation theory with conditions equal-strain and axisymmetric given by Barron.

7.7.3 Stability of embankment checking shall meet the following requirements:

1 For stability checking of embankment on soft ground, the consolidation effective stress methods in Sweden circle slip method and improved total stress method may be adopted. If conditions permit, the simplified Bishop method and Janbu's simplified method of slice also may be adopted.

2 To check the embankment, a stability safety factor shall be calculated respectively by the load during construction and operation. For the load during construction, only the weight of the embankment is considered. For the load during operation, the weight of the embankment and increased weight caused by pavement and vehicle load shall be considered.

7.7.4 Selection of ground improvement method shall meet the following requirements:

1 The ground improvement method shall be selected based on the soft soil thickness, the height of the embankment, the requirements of stability and post-construction settlement, construction machines, material, environment and construction period. The economy, the simplicity, and the availability of local construction material shall also be considered.

2 For ground with very soft soil, complex ground condition, tight construction schedule, lack of construction material and other special requirements, multiple ground improvement methods should be adopted.

7.7.5 Reinforcement measures of subsoil shall meet the following requirements:

1 Permeable horizontal bedding with a thickness of 0.50 m should be set at the bottom of the embankment on soft ground. The width shall be 0.5–1.0 m at both sides. If the bedding is also used as a filter, the thickness shall increase properly.

2 For shallow soft soil with small thickness, the soft soil may be replaced with sand, gravel and crushed stones.

3 Lightweight materials such as fly ash and foam polystyrene (EPS) blocks may be used to fill a lightweight embankment. When the fly ash embankment is adopted, relevant technical requirements of Sect. 3.9 shall be complied with.

4 For reinforced embankment, geosynthetics materials with high intensity, small deformation and anti-aging shall be adopted as embankment reinforcement materials.

5 The loading berm may be set at one side or both sides of the embankment; its height should not exceed half the embankment height and its width shall be determined by a stability calculation.

7.7.6 Drainage consolidation method shall meet the following requirements:

1 Soft ground treatment schemes of sand bedding preload, bagged sand well (prefabricated plastic drain plate preloading) or vacuum-surcharge preloading method may be chosen.

2 Bagged sand well, prefabricated plastic drain plate or various other materials may be selected as the vertical drain according to properties of soft soil, filling materials, construction process, etc. The vertical drain should be arranged in the form of an equilateral triangle; its length is determined according to the stability and deformation of the subsoil. Vertical drains should run through thin soft soil layers. The time period for preloading should be not less than 6 months.

3 Preloading time shall be determined according to the requirements of post-construction settlement or consolidation degree of subsoil. The completed settlement of subsoil during preloading shall not be less than the difference between the settlement of pavement design at the end of working life and allowable post-construction settlement. If necessary, the consolidation of subsoil at the end of preloading shall meet the requirements of embankment stability.

4 Vacuum-surcharge preloading may be used in soft ground treatment in the section of high fill and bridge abutments. When the vacuum-surcharge preloading method is adopted, vertical drain such as sand drain wells and prefabricated plastic drain plates shall be set in the subsoil. The pressure of vacuum behind the seal membrane with vacuum preloading shall not be less than 70 kPa. In cases where the surface layer has a good breathable layer or a permeable layer with sufficient water supply in the treatment range, measures of cutting off the breathable layer and permeable layer shall be adopted.

7.7.7 The design of granular soil piles treatment foundation shall meet the following requirements:

1 Vibroflotation granular soil piles may be used in subsoils with vane shear strength more than 15 kPa while sinking pipe granular soil piles may be used in subsoils with vane shear strength more than 10 kPa.

2 Sand, gravel and crushed stone may be used as material of granular soil piles. The material shall not use single size material and the fine content shall not be greater than 5%.

3 The diameter, depth and spacing of granular soil piles shall be determined according to a stability and settlement analysis. The clear spacing between adjacent piles shall not be more than four times the diameter of a pile.

4 When the global shear safety factor of an embankment on composite ground with granular soil piles is calculated, the shearing strength of the slip surface within the composite ground may adopt the shearing strength of the composite ground τ_{ps}; the strength is calculated according to Eq. (7.7.7.1).

$$\tau_{ps} = \eta\tau_p + (1-\eta)\tau_s \tag{7.7.7.1}$$

where:

η—the area replacement rate of pile and soil;

τ_p—the shearing strength of granular soil pile (kPa);

τ_s—the shearing strength of subsoil (kPa).

5 The settlement of subsoil S_z within the depth of the length of granular soil pile shall be calculated using Eq. (7.7.7.2).

$$S_z = \mu_s S \tag{7.7.7.2}$$

$$\mu_s = \frac{1}{1 + (n - 1)} \tag{7.7.7.3}$$

where:

μ_s—the stress reduction coefficient of soil between piles;

n—the pile to soil stress ratio; it should be determined by testing; if there is no test data, it may be taken as 2–5, when the soil at the bottom of the piles is good and soil between the piles is poor, then the high numerical value can be taken and vice versa;

S—the settlement of the original subsoil within the depth of the length of granular soil pile.

7.7.8 The design of reinforcing soil pile treatment foundation shall meet the following requirements:

1 Deep mixing method may be used to strengthen soft soil foundation with shear strength of vane shear not less than 10 kPa. When using dry jet mixing method, the depth should not be greater than 15 m; when using slurry-grouting method, the depth should not be greater than 20 m.

2 The diameter, depth and spacing of reinforcing soil piles shall be determined according to stability checking and shall meet the requirements of post-construction settlement. The clear spacing of adjacent piles shall not be more than four times the diameter of the pile.

3 When calculating the global shear safety factor of embankment on composite ground with a reinforcing soil pile, the shearing strength of the slip surface within the composite ground shall adopt composite ground shearing strength τ_{ps} and be calculated according to Eq. (7.7.7.1).

4 For shearing strength of the reinforcing soil pile, strength at the age of 90 days is taken as standard strength and it may be calculated as half the unconfined compressive strength q_u of the original specimen drilled in the test section. It also may be obtained by unconfined compressive strength of specimens prepared indoors with design mixture ratio multiplied by with a reduction factor of 0.3,i.e.$\tau_p = 0.3q_u$.

5 The settlement of composite ground with reinforcing soil piles shall be calculated in two parts-settlement of reinforced area with composite ground S_1 and settlement of underlayer of reinforced area S_2.The settlement of reinforced area S_1 should be calculated by the composite compression modulus method and the settlement of

underlayer S_2 may be calculated by the compression modulus method in *Code for Design of Building Foundation* (GB 50,007).

6 Composite compression Modulus E_{ps} shall be calculated according to the Eq. (7.7.8.1).

$$E_{ps} = \eta E_p + (1-\eta)E_s \tag{7.7.8.1}$$

where:

E_p—compression modulus of pile (MPa);

E_s—compression modulus of soil (MPa).

7.7.9 The design of cement fly ash gravel (CFG) piles treatment foundation shall meet the following requirements:

1 CFG pile may be used to improve the soft ground with vane shear strength less than 20 kPa.

2 Gravel and crushed stone may be used as material for CFG piles. When pumping, the maximum size should be not greater than 25 mm and 20 mm for gravel and crushed stone, respectively. When vibrated grouting, the maximum size should not be greater than 50 mm. Sand and fines may be used to improve the gradation. 32.5 Portland cement and classes II and III fly ash should be adopted.

3 Gradation of the CFG material shall be determined based on the slumps and the design strength of the CFG pile. The design strength of CFG pile shall use the unconfined compression strength at 28 days.

4 The strength of CFG pile should be 5–20 MPa which shall meet the requirements of the settlement and stability. The strength of CFG piles used to support structures shall also meet the requirements of bearing capacity.

5 The pile diameter, length and spacing shall be determine based on the bearing capacity, the settlement, soils and machine condition. The pile toe shall be located on good soil layers and the maximum pile length should not be larger than 30 m. The pile spacing should be 4–5 times pile diameter.

6 The cushion thickness for CFG piles should be 0.3–0.5 m. When CFG piles have large diameter or pile spacing, a thicker cushion should be used. Medium sand, coarse sand, well-graded gravel or crushed stones should be used the cushion material and the maximum particle size should not be larger than 30 mm.

7 The settlement and stability calculation of CFG composite foundation shall follow Clause 7.7.7.

7.7.10 The design of dynamic compaction and dynamic compaction replacement treatment foundation shall meet the following requirements:

1 The dynamic compaction method may be adopted when there are many layers of silt in saturated soft clay subsoil or when aggregates such as block stone, crushed gravel, cobble stone and so forth are used to backfill by heavy tamping replacement.

2 The treatment depth shall be determined based on the soil condition which should penetrate soft soil layer and reach hard soil layer except thick silt layer. The treatment depth should not be greater than 7 m.

3 Before dynamic compaction construction, representative sections shall be selected for trial ramming in the construction field to guide construction in large areas.

4 Effective reinforcing depth of dynamic compaction d shall be determined according to field trial ramming or local experience, also may be estimated according to Eq. (7.7.10.1).

$$d = \alpha\sqrt{mh} \tag{7.7.10.1}$$

where:

m—the mass of the rammer (t);

h—the falling distance of rammer (m);

α—the correction factor, which is related to factors such as the condition of soil, underground water level, size of tamping energy, area of rammer base and so on; the value ranges from 0.34–0.80 and it shall be determined according to the result of trial ramming in construction field.

5 The tamping number of tamping points (in the best of tamping energy) shall be determined according to field trial ramming and shall meet the following conditions: The average ramming settlement of the last two hits shall meet the requirement in Table 7.4.The uplift of surrounding soil shall be limited and the depth of carter shall also be limited to eliminate the difficulty of lifting up hammers.

6 Tamping number at a single tamping point shall be determined based on the field trail. The replacement shall penetrate the soft soil to reach the design depth. The settlement of each tamping shall not cause difficulty of lifting up the hammer. The accumulated settlement should be 1.5–2.0 times of pile length. The average settlement of last two hits shall meet the requirement in Item 5 in this clause.

7 Tamping points may be arranged in the form of squares or equilateral triangles and should have spacing of 5–7 m.

8 Replaced pile spacing shall be determined based on the load magnitude and the bearing capacity of the native soil. The pile spacing may be 2–3 times of the hammer diameter when the damping is fully arranged. For a single footing or stripe footing, the pile spacing may be 1.5–2.0 times of the hammer diameter. The pile diameter may be determined as 1.1–1.2 times of the hammer diameter.

9 A cushion shall be placed above the replaced piles and the thickness shall be not less than 0.5 m. The same material may be used both for cushion and pile and the particle size should not be greater than 100 mm.

Table 7.4 Average settlement for last two hits

Average E for single hit (kN · m)	Settlement for last two hits (mm)
$E < 2000$	≤ 50
$2000 < E \leq 4000$	≤ 100
$E > 4000$	≤ 200

10 Settlement and stability of dynamic compaction replaced composite foundation shall follow Clause 7.7.7.The stress concentration ratio may use the following values of 2–4 for clay foundation or 1.5–3 for silt and sand foundation.

7.7.11 The design of rigid pile composite foundation treatment foundation shall meet the following requirements:

1 Rigid piles may be used to treat thick soft soil with heavy load and strict settlement requirement in high embankment, the connection of embankment with bridge and channel and widened embankment.

2 Pile cap should be used for rigid piles along with geosynthetic reinforced cushion.

3 The layout of rigid piles may be rectangular or regular triangular. The diameter, length and pile spacing of rigid piles shall be determined based on settlement calculation. The pile spacing should not be less than 5 times of pile diameter.

4 The pile cap may use cylinder, platform or back taper shapes. The pile cap size should be 1.0–1.5 m in diameter and 0.3–0.4 m in thickness.

5 The settlement of composite foundation with rigid piles shall be calculated using 1-D layered-wise summation method using Eq. (7.7.11) and may be without consideration of soil compression between piles:

$$S = \psi_{\mathrm{P}} \sum_{j=1}^{m} \sum_{i=1}^{n_j} \frac{\sigma_{j,i} \Delta h_{j,i}}{E_{sj,i}} \tag{7.7.11}$$

where:

S—final settlement (m);

m—number of layers below the pile toe;

$E_{sj,i}$—the modulus of the soil of sublayer i in layer j (MPa);

n_j—number of layers for layer j;

$\Delta h_{j,i}$—thickness of sublayer i in layer j (m);

$\sigma_{j,i}$— vertical additional stress on sublayer i in layer j (kPa),which may be determined based on Appendix E in *Code for Design of Building Foundation* (GB 50,007);

ψ_{P}—empirical coefficient for pile settlement calculation which shall be determined based on the local experience.

6 Stability of composite foundation with rigid piles uses circular sliding method. The shear strength on the sliding surface uses the combined shear strength of pile and soil calculated using Eq. (7.7.7.1).The shear strength of pile may use half of its unconfined compression strength at 28 d.

7.7.12 The design of cross section of embankment over soft soil shall consider the settlement, the sunken of the top surface, the width change at bottom and top, and the slope change.

7.7.13 Observation design of settlement and stability shall meet the following requirements:

1 For high fill embankment and bridge abutment embankment on soft ground, design for observation of settlement and stability shall be carried out. The design

items shall include: the measurement positions of settlement observation and lateral displacement (stability),selection and layout of the observation apparatus, observation method and observation frequency. If necessary, deep displacement observation of soft ground shall be carried out.

2 The filling pace of embankment shall meet the following requirements:

(1) Fill time shall not be less than the consolidation time required by subsoil shearing strength increase.
(2) Settlement of the embankment center shall not be larger than 10 mm to 15 mm for every day and night while the displacement of the side pile shall not be larger than 5 mm for every day and night.

7.7.14 For embankment on soft ground, representative sections should be selected to fill the test embankment in advance according to engineering practices.

7.7.15 Pavement paving must be carried out after the settlement is stable and meets the following standards: when the post-construction settlement by calculation is less than the allowable design value and monthly observation settlement in consecutive two months is not more than 5 mm, then excavated road trough may be unloaded and pavement paving may begin.

7.8 Substructure in Red Clay and High Liquid Limit Soil Area

7.8.1 Design of the substructure in red clay and high liquid limit soil area shall meet the following requirements:

1 When route passes through the red clay or high liquid limit soil areas, data such as the distribution and genetic type of red clay or high liquid limit soil, characteristics of soil structure level, moisture status, vertical zoning, distribution of cracks in soil, underground water distribution, physical and mechanical properties, swell-shrink characteristics and so forth shall be identified.

2 Red clay may be classified based on the relationship of ratio of liquid and plastic limits with the limit of the ratio of liquid and plastic limits and soaking characteristics, which may follow Table 7.5. The ratio of liquid and plastic limits and the limit of the ratio of liquid and plastic limits may be calculated using Eqs. (7.8.1.1) and (7.8.1.2).

$$I_r = \frac{w_L}{w_p} \tag{7.8.1.1}$$

$$I'_r = 1.4 + 0.0066w_L \tag{7.8.1.2}$$

where:

I_r—ratio of liquid and plastic limits;

I'_r— the limit of the ratio of liquid and plastic limits;

Table 7.5 Classification of repeat immersion properties of red clay

Classification	Relation between I_r nd I'_r	Characteristics of repeat soaking
I	$I_r \geq I'_r$	Expansion after contraction, able to restore to the in situ
II	$I_r < I'_r$	Expansion after contraction, unable to restore to the in situ

w_L—liquid limit (%);
w_P—plastic limit (%).

3 For red clay and high liquid limit soil with expansibility, the design shall be carried out according to expansive soil substructure.

4 High embankment and deep cutting should be avoided in the substructure design. If high fill or deep cutting cannot be avoided, the substructure design should be determined by comprehensive comparison with the bridge and tunnel options.

5 The design of substructure with red clay and high liquid limit shall consider the influence of climate and water on the behavior of the substructure. The drainage and temperature control for the substructure shall design well with continuous construction and timely close.

7.8.2 The red clay and high liquid limit soil shall not be used as backfill material for substructure. The red clay with compressibility greater than 0.5 MPa^{-1} shall not be used as backfill material for embankment.

7.8.3 The following requirements shall be met when the red clay and high liquid limit soil is used as backfill material:

1 Resilient modulus and CBR of red clay and high liquid limit soil shall meet the requirement in Sects. 3.2 and 3.3.

2 The red clay and high liquid limit soil treated physically may be used as backfill material below substructure, however, it shall not be soaked.

3 The substructure and low embankment backfilled with red clay and high liquid limit soil, inorganic binder shall be added.

4 The optimum water content, maximum dry density and CBR value of the backfill material shall be determined using the modified proctor compaction method. The number of blows shall be adjusted based on the water content for CBR test.

7.8.4 Fill substructure design shall meet requirements as:

1 The design of the substructure fill shall be based on the climate, hydrology, embankment height, properties of red clay and high liquid ratio soil and the treatment method. Red clay and high liquid ratio soil should not be used for steep slopes.

2 Embankment filling height with red clay with physical treatment should not be larger than 10 m. A separation cushion using gravels or crushed stones shall be placed at the embankment bottom, and the thickness should not less than 0.5 m. A geosynthetic filter layer should be placed at the top of the cushion.

3 The side slope of embankment with the height no larger than 10 m should be 1:1.5 to 1:2.When the height of the side slope is larger than 6 m, the slope platform with width no less than 2 m should be set. When side slope height is larger than 10 m, the embankment cross section, side slope grade and embankment protection and

Table 7.6 Side cut slope ratio

Slope height (m)	Side slope rate	Slope platform width (m)
<6	1:1.25–1:1.5	–
6–10	1:1.25–1:1.5	2.0
10–20	1:1.5–1:1.75	≥2.0

reinforcement measures shall be determined by an embankment stability analysis and compliance with the relevant requirements of Sect. 3.6.

4 Embankment side slope protection with modification treatment or non-red clay (high liquid limit soil) outside the seal layer may be treated as general substructure protection.

5 The continuous construction of the embankment should be scheduled in dry season. The top surface shall be covered timely if continuous construction could not be fulfilled.

7.8.5 Excavation substructure design shall meet the following requirements:

1 In the design of excavation substructure, attention shall be paid to the adverse effects of crack development caused by the dry shrinkage of excavation face of Class-1 red clay soil under repeat soaking and soil property changes as a result of repeat soaking. The side slope stability analysis calculation should adopt strength indices such as a saturated shear test and repeat slow shearing test.

2 The height of the side cut slope should not be more than 20 m. The design of cutting slope shall follow the principle of gentle side slope, wide platform and reinforced slope toe. The rate of side slope and platform width may be determined using Table 7.6.When the side slope height exceeds 6 m, a stepped cross section should be adopted in the excavation substructure. If the terrain allows, the side slope should be more gentle.

3 A good drainage system for surface water and groundwater shall be installed for cutting slope. Supporting sewers should be constructed on the surface of the cutting slope and sewers shall be installed under the side ditch. When the presence of the collected water on the slope surface, oblique drainage holes should be constructed.

4 The skeleton plant protection should be adopted for sloping surface protection. When the side slope does not have adequate stability, retaining engineering shall be added. For full closed masonry protection, a drainage cushion layer with thickness of 0.15–0.30 m shall be set at the back of the wall. Drainage holes should be constructed with a spacing of 2.5–3.0 m and a filter layer is also needed.

5 Plants and covers outside of the cutting slope top should be maintained. Blocking water wall or catch water ditch should be constructed at the slope top.

6 Red clay and high liquid limit soil in the excavation of substructure should be over-excavated and replaced or treated using inorganic binders based on the engineering properties of the red clay and the high liquid limit soil. The replacing material should use good permeable materials including gravels and crushed stones.

7 The stalacto-stalagmite and stalagmite in the excavation should be removed. When over wet soil exists between the stalacto-stalagmite and stalagmite with the

natural water content of 5% higher than the plastic limit, the over wet soil in the range of substructure should be replaced with rubbles.

8 The replacement in the substructure should be conducted timely in non-fill and cutting substructure sections. If the construction could not be executed timely, a protection layer with a thickness of 0.3 m should be placed above the elevation of the substructure bottom.

7.9 Substructure in Expansive Soil Area

7.9.1 The substructure in expansive soil area shall meet the following requirements:

1 The following shall be identified for substructure design in expansive soil area: Distribution of expansive soil, genetic type, soil structure level, distribution and buried condition of underground water, mineral composition of expansive soil, physical and mechanical properties, swelling characteristics and so forth.

2 The alignment shall be designed based on the engineering properties of the expansive soil and the highway class as well as consideration of the landform, geology, hydrology and backfill material. The major index shall be determined based on the overall analysis and the comparison of the alignment design.

3 Large filling and large cutting shall be avoided, while shallow cut and low embankment should be preferable in substructure design. When substructure filling or cutting is large, construction is difficult and stability is poor. Comparisons shall be made with bridge and tunnel options. If the substructure is selected to pass through, measures shall be taken to ensure stability.

4 The improvement design shall be determined based on the laboratory test and economic comparison when the expansive soil is used as backfill material. The optimum content of the binder and the soil strength after treatment shall also be determined.

5 When a highway passes through expansive soil section, the substructure design shall mainly be based on the principles of moisture retention, being waterproof and anti-weathering, in combination with slope surface protection, decreasing side slope height, continuous construction, and covering substructure and sloping surfaces in a timely manner.

7.9.2 The expansive soil shall be evaluated based on the landform, soil color, soil structure, geological condition, natural geology and free swelling ratio of soil. The expansive soil can be classified based on the standard absorption moisture. When the standard absorption moisture is higher than 2.5%, the soil shall be expansive soil. The class of expansive soil and the expansive potential shall be determined based on the current *Code for Highway Engineering Geological Investigation* (JTG C20).

7.9.3 Calculation of the expansive soil foundation settlement shall meet the following requirements:

1 When the foundation of earth structures for example retaining wall and the bed of the low highway embankment are expansive soil foundation, the settlement of the expansive soil may be calculated based on Eq. (7.9.3.1) or Eq. (7.9.3.2).

2 The settlement of expansive soil foundation may be calculated as Eq. (7.9.3.1) if the consolidation test results are available:

$$\rho = \sum_{i=1}^{n} \frac{C_s z_i}{(1+e_0)} \lg\left(\frac{\sigma'_f}{\sigma'_{sc}}\right)_i \tag{7.9.3.1}$$

where:

ρ—settlement (mm);
e_0—initial void ratio;
σ'_{sc}—calibrated expansive pressure from constant volume test (kPa);
σ'_f—final effective stress (kPa);
C_s—expansive index;
z_i—original thickness of the soil layer i (mm).

3 The settlement of expansive soil foundation may be calculated as Eq. (7.9.3.2) if the shrinkage test results are available:

$$\rho = \sum_{i=1}^{n} \Delta z_i = \sum_{i=1}^{n} \frac{c_w \Delta w_i}{(1+e_0)_i} \tag{7.9.3.2}$$

$$C_w = \frac{\Delta e_i}{\Delta w_i} \tag{7.9.3.3}$$

where:

C_w—volumetric shrinkage index of expansive soil;
Δe_i—void ratio changes of soil layer i;
Δw_i—moisture content change of soil layer i.

7.9.4 Design of the expansive soil foundations shall use the settlement as the standard (as shown in Table 7.7) to choose the improvement method and the treatment depth.

7.9.5 When the expansive soil is used as backfill material, the expansive soil shall be classified based on the total rate of swelling-shrinkage as shown in Table 7.8 to determine the improvement method and the treatment depth.

7.9.6 Fill substructure design shall meet the following requirements:

1 The highway embankment with expansive soil shall be designed based on the embankment height, the expansive soil type and the improvement method. The prevention of seepage, drainage and moisture maintenance shall be designed well to ensure the stability of the embankment.

2 When the expansive soil is used as backfill material for the substructure and embankment with a height less than 1.5 m, the design shall follow the requirements in Sect. 3.2. When the weak and medium expansive soils are used as backfill material, modification shall be adopted to minimize the total rate of swelling-shrinkage to 0.7%.

3 An expansive soil cushion improved with gravel, crushed stone and modification should be placed at the bottom of the substructure for improved expansive

Table 7.7 Classification of expansive soil foundation

Foundation class	Settlement ρ (mm)	Ground improvement
I	$\rho \geq 200$	Deep foundation should be used for small scale structures. When the embankment height is less than 1.5 m, the expansive soil should be replaced using non-expansive soil or modified soil, and the treatment depth should not be less than 2.0 m
II	$100 \leq \rho < 200$	Shallow foundation may be used for small scale structures and the embedment depth of the foundation should not be less than 1.5 m. The moisture of the soil should be maintained. When the embankment height is less than 1.5 m, the expansive soil need to be replaced using non-expansive soil or modified soil, and the treatment depth should not be less than 1.5 m
III	$40 \leq \rho < 100$	Shallow foundation may be used for small scale structures and the embedment depth of the foundation should not be less than 1.0 m. The moisture of the soil should be maintained. When the embankment height is less than 1.5 m, the expansive soil need to be replaced using non-expansive soil or modified soil, and the treatment depth should not be less than 1.0 m
IV	$15 \leq \rho < 40$	Shallow foundation may be used for small scale structures. When the embankment height is less than 1.5 m, the expansive soil need to be replaced using non-expansive soil or modified soil, and the treatment depth should not be less than 0.5 m
V	$\rho < 15$	Treatment may be not necessary

Table 7.8 Classification of the expansive soil used as backfill material

Backfill class	Total rate of swelling-shrinkage under loading (%)	Usage
Non-expansive	$e_{ps} < 0.7$	May be used directly
Weak-expansive	$0.7 \leq e_{ps} < 2.5$	May be used as backfill material after treatment with coverage, geosynthetic reinforcement and placement of cushion. The expansive soil can be used as backfill material for the substructure after modification
Medium-expansive	$2.5 \leq e_{ps} < 5.0$	The expansive soil may be used as backfill material for the embankment after modification
Strong-expansive	$e_{ps} \geq 5.0$	The expansive soil shall not be used as backfill material for embankment

Note 1. When the embankment height is not less than 3.0 m, the total rate of swelling-shrinkage shall be determined with a loading of 50 kPa

2. When the embankment height is less than 3.0 m, the total rate of swelling-shrinkage shall be determined with a loading of 30 kPa

soil embankment. The cushion should not be less than 0.5 m. The treated expansive soil should be covered by the non-expansive or modified expansive soils with the coverage thickness not less than 2.5 m. When the gravel and crushed stone are used as filter in the coverage layer, a seepage proof layer shall be placed and it may use geomembrane or other materials.

4 A seepage proof layer shall be placed at the bottom of the substructure when gravel and gravelly soil are used as backfill material above the expansive soil foundation. The material including non-expansive soil, modified soil or geomembrane may be used for the seepage proof layer.

5 A modified proctor compaction test on wet soil should be used to determine the optimum moisture content and maximum dry density when the modified expansive soil used as backfill material. When the inorganic binder is used to improve expansive soil, the optimum water content plus 2% determined from modified proctor compaction test on dry soil may be used as standard.

6 When weak expansive soil and medium expansive soil are used to fill embankment, the side slope shall be comprehensively determined according to embankment height, remolded filling properties, and regional climate characteristics, with reference to mature experience of the existing substructure. For a side slope that is not higher than 10 m, the side slope rate and platform setting may be determined using Table 7.9.

7 Embankment side slope protection may be determined by Table 7.10 based on engineering geological conditions and the filling height.

7.9.7 Excavation substructure design shall meet the following requirements:

1 The rate of the expansive soil side cutting slope shall be comprehensively determined based on the properties of soil, weak layers, crack combination relation, climatic characteristic, hydrogeological conditions, natural side slope and stable grade of artificial side slope and so forth.

2 The side slope design shall comply with the principles of gentle side slope, wide platform and reinforced slope toe. The side slope rate and platform width shall be

Table 7.9 Side slope rate and platform width of expansive soil embankment

Expansibility slope height (m)	Side slope rate		Slope platform width (m)	
	Weak expansive soil	Medium expansive soil	Weak expansive soil	Medium expansive soil
<6	1:1.5	1:1.5–1:1.75	May not be set	
6–10	1:1.75	1:1.75–1:2.0	2.0	≥2.0

Table 7.10 Side slope protection measures of expansive soil embankment

Slope height (m)	Weak expansive soil	Medium expansive soil
≤6	Plants	Framework plant
>6	Vegetation protection, framework plant	Supporting seeping groove and arch framework plant

designed in accordance with Table 7.11. The side slope shall be individually designed when its height is more than 10 m. The slope stability shall follow the requirements in Table 3.15. If necessary, the side slope should compare with the tunnel option.

3 Expansive soil within non-filling and cutting substructure shall be over-excavated and the replacement filling with non-expansive and modified soils which may meet the requirements specified in Sect. 3.2. When the permeable material is used as replacing material, a seepage proof layer shall be placed at the bottom of the substructure. The replacement depth of substructure should be increased to 1.0–1.5 m when cutting in strong expansive soils.

4 Side cutting slope protection and reinforcement types may be determined according to geological conditions, environmental factors and height of side slope in Tables 7.12 and 7.13.The side slope shall be protected with closed protection promptly after excavation.

5 When the side slope using plant protection, hardwood species shall not be used. When using masonry protection, a buffer layer with a minimum thickness of 0.5 m shall be set at the back of the wall. The embedment depth of the retaining structures

Table 7.11 Side slope rate and platform width of expansive soil

Type of expansive soil	Slope height (m)	Side slope rate	Slope platform width (m)	Rockfall berm width (m)
Weak expansive soil	<6	1:1.5	–	1.0
	6–10	1:1.5–1:2.0	2.0	1.5–2.0
Medium expansive soil	<6	1:1.5–1:1.75	–	1.0–2.0
	6–10	1:1.75–1:2.0	2.0	2.0
Strong expansive soil	<6	1:1.75–1:2.0	–	2.0
	6–10	1:2.0–1:2.5	≥2.0	≥2.0

Table 7.12 Side slope protection measures of expansive soil cutting

Slope height (m)	Weak expansive soil	Medium expansive soil
≤6	Plant	Framework plant
>6	Framework plant, plant protection, mortar rubble masonry sewer	Arch framework plant, supporting sewer and arch framework plant

Table 7.13 Side slope retaining measures of expansive soil cutting

Slope height (m)	Weak expansive soil	Medium expansive soil	Strong expansive soil
≤6	No setting	Slope toe wall	Curtain wall, retaining wall
>6	Revetment wall, retaining wall	Retaining wall, slide-resistant pile	Pile foundation retaining wall, slide-resistance pile, slope anchor

shall be greater than the influence depth of the climate and the minimum thickness for the seepage proof layer is 0.5 m.

6 Protection of the side slope may use the coverage of the non-expansive clay or the flexible supporting structures when there is no bad interface between the expansive soil slopes with the underlying soil layers. The thickness of the slope coverage layer should not be greater than 2.5 m and the roller should be used for compaction. Drainage cushion and ditches shall be constructed between the coverage layer and the underlying expansive soils. When there is bad interface in side slope, the retaining structures shall be set up based on the slope stability.

7.9.8 Design of the drainage of the expansive soil substructure shall follow the principle of the combined protection with drainage and meet the following requirements:

1 For non-fill and low-fill substructures, a water blocking ditch shall be installed when the highway is lower than the surrounding area.

2 Seepage proof layer and drainage cushion shall be placed at the bottom of the embankment when the groundwater level is high in low embankment sections. If necessary, sewers may be set up at two sides of the substructure.

3 Supported ditches or oblique drainage holes should be set on the side slopes when the soil is wet and the groundwater exists in the cutting substructures. Drainage ditches in the longitudinal direction should be set the side ditches. Transversal drainage ditches shall be constructed at the interface of backfilling and cutting sections.

4 Expansive soils at 3–5 m outside form the top of the cutting slope shall be replaced with non-expansive soil or placing seepage proof geomembranes to prevent seepage.

7.10 Substructure in Loess Region

7.10.1 Design of the substructure in loess region shall meet the following requirements:

1 The distribute scope of loess, thickness, variation law, genetic type and stratum characteristics of loess along the line, geomorphic unit, surface drainage and underground water that locate at the road line, loess physical and mechanical properties, collapsibility type and collapse classification in different stratum shall be identified in the design of substructure at a loess region.

2 For a loess highland ridge region, substructure shall avoid the edge of highland ridge and sloping areas with landslide, collapse, trapped-hole group, gully development and exposed underground water. If substructure passes through those areas, there shall be adequate bases and practical and reliable engineering measures.

3 For a gully head and trapped hole located near substructure, their development trend and extent of damage to the substructure shall be analyzed and their effect on substructure stability shall also be considered during design.

4 The location, shape, development and water resource and quantity of the collapsed loess holes close to the alignment shall be investigated which can be used to evaluate the damage extent of the collapsed holes to the substructure.

5 In a collapsible loess area, the substructure should be set at the section with slight collapsible classification, a thin collapsible soil layer and good drainage conditions.

6 The design of substructure in loess area shall pay special attention to strengthening drainage and adopting interception and dispersion principles. Scour prevention, anti-seepage, comprehensive drainage and protection systems that are beneficial to soil and water conservation shall be utilized. Mutual interference between the farmland water conservancy facility and the substructure should be properly treated.

7.10.2 Fill Substructure design shall meet the following requirements:

1 When fill substructure is to be built at a loess area, the strength of filling, compaction and treatment of the base shall comply with the requirements of Sect. 3.2. If not, the loess shall be treated using inorganic binders.

2 The substructure in loess area shall be designed based on the highway class, the side slop height and the engineering properties of the foundation soil, and the slope stability also shall be considered to determine the type of embankment slope and slope angle. When the embankment subsoil is in good condition or is being treated and slope height is not more than 30 m, the embankment cross section and the side slope may be selected according to Table 7.14. The stepped cross section should be set for regions with an average annual rainfall more than 500 mm.

3 If the height of the embankment side slope is more than 30 m, the design shall be compared with bridge options and individual special design shall be done in accordance with the requirements of Sect. 3.6.

4 The side slope stability calculation should adopt arc method and the stability factor shall in accordance with the required value in Sect. 3.6. The shear strength of filling shall be determined according to the requirements of designed compaction by the quick shear test after compaction.

5 For an embankment with a height greater than 20 m, the widening value of the substructure top shall be preset according to post-construction settlement; post-construction settlement may be estimated by 0.7 and 1.5% of the embankment height.

6 A higher compaction may be used when the embankment height is greater than 20 m.

7 The treatment of saturated loess and the new loess foundation with low bearing capacity may follow the requirements in Sect. 7.7.

7.10.3 Excavation substructure design shall meet the following requirements:

Table 7.14 Embankment cross section and side slope rate

Cross section	Rate of side slope subsection under substructure		
	$0 < H \leq 10$ m	$10 < H \leq 20$ m	$20 < H \leq 30$ m
Broken line shape	1:1.5	1:1.75	1:2
Ladder shape	1:1.5	1:1.75	1:1.75

1 The form of loess side cutting slope shall be determined according to Table 7.15 on the basis of loess category, loess uniformity and height of the side slope. The cutting slope of Motorway and Class-1 highway should adopt step-shape. The width of the small side slope platform should be 2.0–2.5 m. The width of the large side slope platform should be determined by stability analysis and should range from 4 to 6 m. For an area with more than 250 mm of average annual rainfall, an intercepting ditch shall be set on the platform and the ditch shall be protected.

2 When the height of excavation side slope is no more than 30 m, the rate of the side slope shall be determined based on the geomorphic unit, causes of times, tectonic joint, underground water distribution, rainfall, height of side slope and construction method, in combination with the natural or artificial stable height of the side slope according to Table 7.16.

3 When the loess side cutting slope height is more than 30 m, it shall be compared with the tunnel option. High cutting side slope shall be individually designed according to the requirements in Sect. 3.7.

4 For deep cut substructure with a large platform, the stability shall be checked not only for the entire height of the side slope but also for the upper and lower side slopes locally adjacent to the large platform.

5 Side slope protection shall be selected according to the requirements in Chap. 5 on the basis of soil type, amount of rainfall, climatic conditions, height and grade of side slope, source of side slope protection materials and so forth. Plant protection should be used for cutting slopes and the supported structures shall be used for slopes with low factor of safety.

6 Necessary protection works and measures such as intercepting and draining underground and surface water and preventing the surface water from leakage shall be carried out at cut sections with groundwater activity.

7.10.4 The collapsible loess soil foundation shall be designed including identifying the type of collapse, calculating the settlement of the foundation and determining the collapse class of the foundation. The design shall meet the following requirements:

Table 7.15 Cutting side slope form and its applicable conditions

Side slope form	Application conditions
Line (only one grade of side slope)	(1) Homogeneous soil, Q_4 and Q_3 loess side slope height $H \leq 15$ m; Q_2 and Q_1 loess side slope height $H \leq 20$ m (2) Non-homogeneous soil, side slope height $H \leq 10$ m
Broken line (upper gentle and lower steep)	Non-homogeneous soil, side slope height $H \leq 15$ m
Step-shaped	(1) Homogeneous soil, Q_4 and Q_3 loess side slope height of 15 m $< H \leq 30$ m; Q_2 and Q_1 loess side slope height 20 m $< H \leq 30$ m (2) Non-homogeneous soil, side slope height 15 m $< H \leq 30$ m

Table 7.16 Side cutting slope form and its applicable conditions

Zoning	Category		Slope height (m)			
			≤6	6–12	12–20	20–30
I Southeast	New loess Q_3,Q_4	Side slope deposits	1:0.5	1:0.5–1:0.75	1:0.75–1:1.0	–
		Diluvial	1:0.2–1:0.3	1:0.3–1:0.5	1:0.5–1:0.75	1:0.75–1:1.0
	New loess Q_3		1:0.3–1:0.5	1:0.4–1:0.6	1:0.6–1:0.75	1:0.75–1:1.0
	Old loess Q_2		1:0.1–1:0.3	1:0.2–1:0.4	1:0.3–1:0.5	1:0.5–1:0.75
II Central district	New loess Q_3,Q_4	Side slope deposits	1:0.5	1:0.5–1:0.75	1:0.75–1:1.0	–
		Diluvial, alluvial	1:0.2–1:0.3	1:0.3–1:0.5	1:0.5–1:0.75	1:0.75–1:1.0
	New loess Q_3		1:0.3–1:0.4	1:0.4–1:0.5	1:0.5–1:0.75	1:0.75–1:1.0
	Old loess Q_2		1:0.1–1:0.3	1:0.2–1:0.4	1:0.3–1:0.5	1:0.5–1:0.75
	Red loess Q_1		1:0.1–1:0.2	1:0.2–1:0.3	1:0.3–1:0.4	1:0.4–1:0.6
III Western	New loess Q_3,Q_4	Side slope deposits	1:0.5–1:0.75	1:0.75–1:1.0	1:1.0–1:1.25	–
		Diluvial, alluvial	1:0.2–1:0.4	1:0.4–1:0.6	1:0.6–1:0.75	1:0.75–1:1.0
	New loess Q_3		1:0.4–1:0.5	1:0.5–1:0.75	1:0.75–1:1.0	1:1.0–1:1.25
	Old loess Q_2		1:0.1–1:0.3	1:0.2–1:0.4	1:0.3–1:0.5	1:0.5–1:0.75
IV Northern	New loess Q_3,Q_4	Side slope deposits	1:0.5–1:0.75	1:0.75–1:1.0	1:1.0–1:1.25	–
		Diluvial, alluvial	1:0.2–1:0.4	1:0.4–1:0.6	1:0.6–1:0.75	1:0.75–1:1.0
	New loess Q_3		1:0.3–1:0.5	1:0.5–1:0.6	1:0.6–1:0.75	1:0.75–1:1.0
	Old loess Q_2		1:0.1–1:0.3	1:0.2–1:0.4	1:0.3–1:0.5	1:0.5–1:0.75
	Red loess Q_1		1:0.1–1:0.2	1:0.2–1:0.3	1:0.3–1:0.4	1:0.4–1:0.6

Note Side slope values in the table are the average values after the platform is set

1 The collapse types of the loess region shall be judged by measured deadweight collapse settlement or accumulative calculated deadweight collapse settlement through the laboratory compression test. When collapsible loess with measured or calculated deadweight collapse settlement is not more than 70 mm, it shall be defined as a non-deadweight collapsibility loess field; when collapsibility loess with measured or calculated deadweight collapse settlement is more than 70 mm, it shall be defined as a deadweight collapsibility loess field.

2 The collapse by deadweight of the collapsible loess shall be calculated based on Eq. (7.10.4.1):

$$\Delta_{zs} = \beta_0 \sum_{i=1}^{n} \delta_{zsi} h_i \tag{7.10.4.1}$$

where:

Δ_{zs}—deadweight collapse settlement in collapsible loess field (mm);

δ_{zsi}—deadweight collapse coefficient in layer i soil;

h_i—thickness of layer i (mm);

β_0—correction factor based on the local soil condition. If not available, use 1.50 for Longxi area; 1.2 for Longdong-shanbei-jinxi area;0.9 for Guanzhong area and 0.50 for other areas.

3 The settlement of the collapsible loess foundation shall be calculated using Eq. (7.10.4.2):

$$\Delta_s = \beta_0 \sum_{i=1}^{n} \beta \delta_{si} h_i \tag{7.10.4.2}$$

where:

Δ_s—total collapse settlement (mm);

δ_{si}—collapse coefficient of layer i;

β—correction factor considering the water intrusion and lateral expansion; if not available, $\beta = 1.5$ for the depth of 0–5 m below the foundation;$\beta = 1.0$ for the depth of 5–10 m below the foundation.

4 Collapse classification of the collapsible loess subsoil shall be judged according to Table 7.17 on the basis of accumulative overall collapse settlement and calculated deadweight collapse settlement of soil layers under the foundation base.

7.10.5 Design of the substructure in collapsible loess area shall meet the following requirements:

1 When Motorway and Class-1 highway pass through collapsible loess and higher compression loess sites, the treatment depth of collapsible loess may be determined according to Table 7.18 based on the height of fill embankment, probability of soaking by water, extent of damage due to soaking and difficulty of restoration after soaking damage.

Table 7.17 Collapse classification of collapsible loess subsoil

Collapse type		Non-deadweight collapsibility loess field	Deadweight collapsibility loess filed	
Calculated deadweight collapse settlement Δ_s(mm)		$\Delta_{zs} \le 70$	$70 < \Delta_{zs} \le 350$	$\Delta_{zs} > 350$
Overall collapse settlement Δ_s(mm)	$\Delta_s \le 300$	I (slight)	II (medium)	–
	$300 < \Delta_s \le 700$	II (medium)	II (medium)or III (serious)	III (serious)
	$\Delta_s > 700$	II(medium)	III (serious)	IV (very serious)

Note When $\Delta_s > 600$ mm and $\Delta_{zs} > 300$ mm, it is designated as Grade II; others is designated as Grade II

Table 7.18 Least treatment depth of collapsible loess foundation

Embankment height	Collapse classification and characteristic							
	Frequent running water (high probability of soaking)				Seasonal running water (low probability of soaking)			
	I	II	III	IV	I	II	III	IV
High embankment (>4 m)	2–3	3–5	4–6	6	0.8–1	1–2	2–3	5
Zero fill, excavation substructure and low embankment (≤4 m)	0.8–1	1–1.5	1.5–2	3	0.5–1.0	0.8–1.2	1.2–2.0	2

Note 1. The substructure adjacent to abutment and high retaining wall substructure (height > 6 m) should eliminate all collapsibility of the foundation or penetrate all collapsible soil layers
2. Minimum treatment depth of collapsible loess foundation of excavated substructure is calculated from the top surface of substructure

2 The treatment design of collapsible loess foundation shall be based on highway grade, collapsible grade, treatment depth requirements, construction conditions, material sources and the impact on the surrounding environment, and the treatment measures shall be determined according to Table 7.19 after technical and economic comparison.

3 When the loess foundation may collapse due to farmland irrigation, surface reinforcement and anti-seepage treatment or lateral anti-seepage wall may be applied 5–10 m outside the foot of the slope on both sides of the road.

4 The treatment width of collapsible loess foundation shall conform to the following provisions:

Table 7.19 Commonly used treatment measures for collapsible loess

Treatment measures	Scope of application	Effective reinforcement depth
Replacement cushion method	Above groundwater level, local or whole treatment	1–3 m
Impact roller compaction	I–II grade non.gravity and I grade gravity collapsible loess with the degree of saturation $S_r \leq 60\%$	0.1–1 m, ≤ 1.5 m
Surface re.compaction		1-3 m
Heavy tamping replacement	Above groundwater level, collapsible loess with the degree of saturation $S_r \leq 60\%$	3–6 m, ≤ 8 m
Compaction method (lime.soil, gravel compaction pile)	Above groundwater level, collapsible loess with the degree of saturation $S_r \leq 65\%$	5–12 m, ≤ 15 m
Pile foundation	Used to deal with bridge, culvert, retaining wall and other structural foundation	≤ 30 m

(1) The non-gravity collapsible loess site shall be no less than 1 m outside the foundation bottom, and the self-gravity collapsible loess site shall be no less than 2 m outside the foundation bottom in the retaining wall section.
(2) The embankment section shall be no less than 1 m outside the drainage ditch at the foot of the slope, and the cutting section is the whole excavation surface of the substructure.

5 The loess caves which endanger the stability of substructure shall be treated. The treatment method of loess subsidence shall be determined according to the depth and size of subsidence. The treatment measures of excavation, backfilling, compaction, sand filling and grouting may be adopted. The treatment width depends on the grade of highway. Interception and drainage measures shall be taken for surface water flowing to caves, and filling and compacting for cracks and water-bearing depressions at the top of the cut.

7.10.6 Construction monitoring should be carried out for loess high embankment, deep cutting and collapsible loess foundation treatment, and monitoring design should conform to the requirements of Clauses 3.6.14 and 3.7.11 of this standard.

7.11 Substructure in Saline Soil Area

7.11.1 Design of the substructure in saline soil area shall meet the following requirements:

1 Investigation shall be carried out along the highway to collect rainfall, evaporation, temperature, landform, engineering geology, and hydraulic geology. The distribution range of different types of saline soil, characteristics of salt-containing, underground water and surface drainage conditions shall be determined for highways in the saline soil area to evaluate the bearing capacity, salt expansion, collapsibility and table poly.

2 The location of the substructure shall be selected at place of high elevation, low groundwater level, good drainage condition, low salt content, low mineral content of groundwater, and small distributed area of the saline soil. An embankment shall be needed for the highway.

3 The newly constructing roadbed shall be designed based on the local salt condition, soil behavior, and surface water and groundwater conditions. A comprehensive design is deserved with the considerations of improvement of saline soil, control of backfill material, substructure structure and water proof and drainage measures to ensure the strength and stability of the substructure.

4 The rebuilding substructure shall be designed based on the damage of the substructure and pavement, salt content and type in the backfill material and hydrological geological condition. The design plan shall be determined based on the comparison of the cost of ground improvement and rebuilding plan.

7.11.2 Saline soil may be classified according to Tables 7.20 and 7.21 on the basis of contained salt property.

Table 7.20 Saline soil classification by contained salt property

Name of saline soil	Ratio of ion content	
	Cl^-/SO_4^{2-}	$CO_3^{2-}+HCO_3^-/Cl^-+SO_4^{2-}$
Chlorine saline soil	> 2	–
Chlorite saline soil	1–2	–
Sulfurous acid saline soil	0.3–1.0	–
Sulfuric acid saline soil	< 0.3	–
Carbonate saline soil	–	>0.3

Note The ion content is calculated by mmol number of 1 kg soil (mmol/kg)

Table 7.21 Saline soil classified by salinization

Name of saline soil	Fine-grained soil The average salt content of soil layer (mass percentage)		Coarse-grained soil The average salt content of soil passing 1 mm sieve opening (mass percentage)	
	Chlorine saline soil and chlorite saline soil	Sulfuric acid saline soil and sulphurous acid saline soil	Chlorine saline soil and chlorite saline soil	Sulfuric acid saline soil and sulphurous acid saline soil
Weak saline soil	0.3–<1.0	0.3–<0.5	2.0–<5.0	0.5–<1.5
Medium saline soil	1.0–<5.0	0.5–<2.0	5.0–<8.0	1.5–<3.0
Strong saline soil	5.0–8.0	2.0–5.0	8.0–10.0	3.0–6.0
Over saline soil	>8.0	>5.0	>10.0	>6.0

Note The ion content is calculated by total salt content in 100 g dry soil

7.11.3 The saline soil foundation shall be evaluated based on the salt expansion and collapse behavior of soil. The following requirements shall be met:

1 The salt expansion behavior of soil shall be determined based on the salt expansion degree of the soil at 1 m depth from the ground surface. When the monitoring duration is insufficient, the sodium sulfate content may be used as evaluation index. The salt expansion degree or the sodium sulfate content shall meet the requirements in Table 7.22.

2 When the depth of the groundwater level is less than 3.0 m or surface water always exists in saline soil substructure sections, the collapse settlement shall be calculated based on Eq. (7.11.3) to evaluate the collapsibility of the foundation. The collapse settlement of the foundation for different class highways should follow the specification in Table 7.23.

$$\Delta S = \sum_{i}^{n} \delta_j h_i \tag{7.11.3}$$

Table 7.22 Allowable salt expansion degree

Highway class	Substructure height h (m)	Salt expansion degree η (%)	Sodium sulfate content Z(%)
Motorway, class-1 highway	≤2	≤1	≤0.5
	>2	≤2	≤1.2
Class-2 highway and lower level highway	≤2	≤2	≤1.2
	>2	≤4	≤2.0

Table 7.23 Collapse index of the saline soil foundation

Highway class	Mortorway and class-1 highway	Class-2 highway	Class-3 and 4 highway
Collapse settlement ΔS (mm)	<70	<150	<400

where:
ΔS—collapse settlement of the foundation (mm);
δ_i—collapse coefficient of the soil layer i;
h_i—thickness of the soil layer i (mm);
n—layer numbers in the influence depth of saline soil.

7.11.4 The design of the saline soil subgrade shall follow the following specifications:

1 In the saline soil substructure sections with acceptable salt expansion degree and collapse settlement, the salt rust, salt crest and growing saline-alkaline tolerance plant shall be removed and replaced with gravel. The treatment depth should be 0.3–0.5 m.

2 In the saline soil substructure sections without acceptable salt expansion degree, the following measures may be used: increase the replacement depth; replace soil with non-saline soil; increase the substructure height properly.

3 In the saline soil substructure sections without acceptable collapse settlement, the improvement method may include removing surface soil, impact compaction, soaked, soil replacement and dynamic compaction, and the drainage of the substructure is also needed to design properly.

4 The saline weak soil foundation may be treated using soil replacement, cement stabilized macadam, dynamic compaction replacement and gravel pile. The post-construction settlement of the treated foundation shall meet the requirements in Sect. 7.7.

7.11.5 Embankment should be used for substructure in saline soil areas. When the non-fill or cutting substructure have to be used, the saline soil within the roadbed range shall be over-excavated and replaced with permeable material without salt, and as separation layer is also necessary.

7.11.6 The embankment height in saline soil areas shall be determined based on the type of saline soil, the capillary height, the frost depth, the salt expansive depth and the separation type. When there is no separation layer, the minimum embankment shall be not less than the values in Table 7.24.

7.11.7 Gravel and Aeolian sand should be used as backfill material for saline soil substructures. The usability of saline soil for the embankment filling depends on the highway classification, embankment filling site, local climate characteristics and hydrogeological conditions determined according to Table 7.25.

7.11.8 The embankment side slope in the saline soil area shall be determined according to Table 7.26 on the basis of soil behavior and salinization extent of filling material soil.

7.11.9 A separation layer shall be installed in the embankment in sections with high groundwater level, high capillary and within groundwater influence zone. The design shall meet the following requirements:

1 The elevation of the separation layer shall be 0.2 m higher than the surface water level and meet the requirement for the maximum frost depth. For expressway and newly building class-I highway, the separation layer should be placed below the roadbed.

2 The slope of the separation layer shall be not less than 2%, and the maximum value shall be 5%.

3 The acceptable material for the separation layer may include high permeable gravel and composite seepage-proof geotextile. The thickness of the separation layer should be 0.3–0.5 m, the maximum particle size shall be less than 50 mm, and the silty and clay content shall be not less than 5%.

7.11.10 Drainage system of the saline soil substructure shall use the combined method considering prevention, drainage and dredge. The disease prevention for the saline soil substructure shall follow the specifications including:

Table 7.24 Minimum embankment height in saline soil area

Soil type	Height above ground (m)		Height above underground water level or long-term surface water accumulation level (m)	
	Weak saline soil and medium saline soil	Strong saline soil and over saline soil	Weak saline soil and medium saline soil	Strong saline soil and over saline soil
Gravel soil	0.4	0.6	1.0	1.1
Sandy soil	0.6	1.0	1.3	1.4
Cohesive soil	1.0	1.3	1.8	2.0
Silt	1.3	1.5	2.1	2.3

Note 1. Motorway and Class-1 highway shall be calculated by 1.5–2.0 times the values in above table; Class-2 highway shall be calculated by 1.0 and 1.5 times the value in the table
2. Saline soils with chlorine and chlorite may use lower value

Table 7.25 The usability of saline soil for substructure filling

Soil type	Salt type	Salinization	Motorway,Class-1 highway			Class-2 highway			Class-3 and 4 highway	
			Roadbed	Higher part of embankment	Lower part of embankment	Roadbed	Higher part of embankment	Lower part of embankment	Roadbed	Higher part of embankment
Fine grained soil	Saline soil with chlorine	Weak saline soil	×	○	○	○	○	○	○	○
		Medium saline soil	×	×	○	×	▲[2]	○	×	○
		Strong saline soil	×	×	×	×	×	▲[3]	×	▲[3]
		Over saline soil	×	×	×	×	×	▲[3]	×	×
	Saline soil with Sulfate	Weak saline soil	×	×	○	×	○	○	▲[2]	○
		Medium saline soil	×	×	×	×	×	○	×	▲[2]
		Strong saline soil	×	×	×	×	×	×	×	×
		Over saline soil	×	×	×	×	×	×	×	×
Coarse-grained soils	Saline soil with chlorine	Weak saline soil	▲[1]	○	○	○	○	○	○	○
		Medium saline soil	×	▲[1]▲[2]	○	▲[1]	○	○	○	○
		Strong saline soil	×	×	○	×	▲[3]	○	×	○

(continued)

Table 7.25 (continued)

Soil type	Salt type	Salinization	Motorway,Class-1 highway			Class-2 highway			Class-3 and 4 highway	
			Roadbed	Higher part of embankment	Lower part of embankment	Roadbed	Higher part of embankment	Lower part of embankment	Roadbed	Higher part of embankment
		Over saline soil	×	×	×	×	×	▲[3]	×	▲[3]
Coarse-grained soils	Saline soil with Sulfate	Weak saline soil	▲[1]▲[2]	○	○	▲[1]	○	○	○	○
		Medium saline soil	×	×	○	×	○	○	▲[1]	○
		Strong saline soil	×	×	×	×	×	▲[1]	×	▲[3]
		Over saline soil	×	×	×	×	×	×	×	×

Note 1. ○-useable; ×-unusable
2. ▲[1]-Coarse grained soil may be useable except the fine grained sand or gravel
3. ▲[2]-The saltine soil may be useable in sections without surface water ponding and the groundwater level is not less than 3 m deep
4. ▲[3]-In over-dry area, the usage may be evaluated

Table 7.26 Rate of side slope of saline Soil area embankment

Soil type	Salinization extent of fill	
	Weak saline soil and medium saline soil	Strong saline soil
Gravel soil	1:1.5	1:1.5
Sandy soil	1:1.5	1:1.5-1:1.75
Silty soil	1:1.5-1:1.75	1:1.75-1:2.00
Clay soil	1:1.5-1:1.75	1:1.75-1:2.00

1 Berms should be constructed at both sides of the substructure in the sections with sufficient groundwater and bad hydrological geological conditions. The width of the berm should be not less than 2 m, and the transversal slope angle shall not less than that of the shoulder slope.

2 Drainage ditches may be constructed at one or both side(s) to low the groundwater level or block the drainage from the farmland in sections with high groundwater level or farmland with drainage and irrigation ditches. The distance from the ditch to the substructure toe shall be not less than 2.0 m. If possible, alkali discharge ditches may be constructed, and the distance from the ditch to the toe shall not less than 5.0 m. The ditch depth shall not be less than 1.0 m.

3 Evaporation pond may be set in the difficult drainage sections if possible. The distance from the evaporation pond side to the substructure toe should be greater than 10 m.

7.11.11 Dry salt lake substructure shall be designed based on the following requirements:

1 The types of rock salt, engineering properties, corrosion conditions, underground water level and salt-containing characteristics shall be identified in the design.

2 Rock salt may be used as backfill material for dry salt lake embankment. For Class-III and IV highways, a low embankment substructure cross section may be used but the embankment height should not be less than 0.3 m and a side slope of 1:1.75–1:3 may be used on the embankment.

3 When there is saturated salt water under the salt lake surface, a substructure cross section with drainage ditch and berm should be adopted. The width of the berm shall be more than 2 m.

4 The soil shall be replaced with gravel, aeoline sand, pebbles or salt cover when it has erosions, solution grooves and solution pools.

7.12 Substructure in Perennially Frozen Area

7.12.1 Design of substructure in perennially frozen area shall meet the following requirements:

1 For the design of substructure in a perennially frozen area, the distribution of perennially frozen soil, frozen soil type, upper limit of frozen layer and hydrogeology

along the highway route shall be ascertained. If substructure is built in a section with frozen swamp, ice mound, ice cone and thermo-karst lake (pool),etc., their ranges, scales, causes and development trends shall be investigated in detail. The classification of perennially frozen soils can be found in Appendix J.

2 When the highway passes through a frozen swamp (swamping wetland) or thermo-karst lake (pool) section, embankment should be adopted. The embankment height shall meet the insulation thickness requirements and shall be higher than the sum of swamp water level in warm seasons, capillary water height, harmful frost-heave height and 0.5 m. The embankment height shall also meet the requirement of thickness for the thermal maintenance layer.

3. When conditions permit, pebble soil and crushed gravel soil should be used with high priority as fill. Frozen soil, fine grained soil and humus-rich soil with a plasticity index larger than 12 and liquid limit larger than 32% shall not be used as fill. The fill of the insulation berm should be obtained from local resources and may use peat, turf, carex tato and fine-grained soil.

4 The design of substructure in perennially frozen soil shall be adopted according to frozen soil types and annual average ground temperatures. A design principle of protect frozen soil, control thaw rate or allowable thaw shall be adopted.

5 Substructure in sections of low ice frozen soil or icy frozen soil may be designed according to the general substructure. Special design shall be done for substructure in the sections of rich ice frozen soil, saturated ice frozen soil or ice layer containing soil and the cutting section of ice mound, ice cone, perennially frozen swamp, thermo-karst lake (pool) or cutting with groundwater.

7.12.2 Design of embankment in perennially frozen soil area shall meet the following requirements:

1 When the principle of protection of frozen soil is used in the design. The minimum embankment height shall be determined based on the area, the backfill material type and the temperature in different areas to maintain the depth of the perennially frozen soil.

2 When the design principle of controlling thawing rate and allowing thawing is adopted, the height of embankment should not be less than 1.5 m. It also should not be too high to prevent uneven deformation and cracking of embankment.

3 When the perennially frozen soil thickness is less than or equal to 2 m or the depth of the perennially frozen soil is less than or equal to 4 m, Thaw in soil should be allowable in design.

4 When the perennially frozen soil thickness is higher than 2 m or the depth of the perennially frozen soil is higher than 4 m, the design shall follow the principles of protection or controlling thaw rate.

5 If the embankment height is not sufficient for the frozen soil protection, other measures including thermal insulation material layer, heat stick, pebble and vent pipe may be used to adjust the temperature.

6 The foundation replacement shall be used in filling-excavation transition and low fill sections. The replacement depth and material is determined based on the calculation of the thermal energy.

7 The thermal insulation layer and thermal insulation berm shall be constructed at the substructure and the side slope toe, respectively when the soil at the bottom of the embankment includes ice and the ice layer is thick. The material of the thermal insulation layer may include local moss, turf, tower head grass, peat and clay.

8 The substructure in the unstable permafrost area shall be comprehensively treated by cooling the foundation and setting insulation layer according to the distribution of frozen soil, filling, substructure excavation and ground temperature, and the installation of insulation layer shall be determined according to the thermal calculation.

9 When the thickness of rich cryosol, saturated cryosol and soil-containing ice layer is small and buried shallowly, measures may be taken to remove the permafrost layer after technical and economic comparison.

10 Measures including thermal insulation material layer, heat stick, pebble and vent pipe should be used to adjust the temperature for the perennially frozen soil with high ice content. When the temperature adjustment is not sufficient, a bridge should be constructed.

11 The embankment height shall be designed based on the thaw and compression settlement and the height and width of the embankment shall be reserved based on the post-construction settlement. Geogrid and geocell may be used for high embankment.

7.12.3 Cutting substructure in perennially frozen soil area shall follow the specifications including:

1 The frozen soil in the roadbed shall be replaced with thermal insulation material, and the thickness shall be determined based on the calculation of the thermal energy.

2 When using pebble and gravel as replacement materials, a composite geomembrance impermeable layer shall be set on the ground to prevent the infiltration of surface water, and the top transverse slope of the impermeable layer shall not be less than 4%.

3 The cut slope rate shall be determined according to the distribution of permafrost, slope surface orientation, ice content and ground temperature. The slope should not be steeper than 1:1.75. The cutting slope should be backfilled with clay and compacted and the grass and plant should be used to protect the slope surface.

4 The cross section of the cutting slope top shall be of a wrapping angle type. The elevation of the wrapping angle of cutting slope top should generally be 0.8 m higher than the original ground with a width which shall not be less than 1.0 m. The outside slope should be 1:1.75 and the inside slope shall be the same as the cutting slope.

5 Design of site investigation shall be performed for sections with low ice content and cutting depth greater than 8 m, high rice content and cutting depth greater than 5 m and side slopes within bad geological conditions.

7.12.4 The design of substructure on permafrost with high ice content and low temperature shall meet the following requirements:

1 The frozen soil should be protected in sections with high ice content and low temperature. The height of the substructure shall be greater than the critical substructure height. Industrial thermal insulation material shall be used in substructure when

the height of the substructure could not reach the critical height due to the limitation of the alignment or landform.

2 The thickness of the thermal insulation layer in the substructure shall be calculated using Eq. (7.12.4) with a concept of equivalent thermal resistance; however, the thickness should be greater than 60 mm. The width of the thermal insulation layer should be same with the pavement layer and the depth of the thermal insulation layer shall depend on the strength and the highway class, which should be 0.30–0.35 m from the top surface of the substructure.

$$d_x = K\frac{d_s \cdot \lambda_e}{\lambda_5} \tag{7.12.4}$$

where:

d_x, d_s—thickness of thermal insulation plate and the equivalent soil layer (mm);

λ_e, λ_s—thermal conductivity coefficient of the thermal insulation plate and the equivalent soil layer;

K—factor of safety for the thermal insulation plate. K need to be 1.5–2.0 for substructure and 1.2–1.5 for substructure side slope.

3 The thermal insulation material in the substructure shall have sufficient thermal resistance and strength. The thermal conductivity should be not less than 0.029 W/(m · K),the water absorption degree should be not less than 0.5%,and the compressive strength should be greater than 600 kPa.

7.12.5 The design of substructure on permafrost with high ice content and high temperature shall meet the following requirements:

1 The principle of controlling thaw rate should be adopted in the frozen soil with high rice content and high temperature. When the substructure is higher than 3.0 m, the pebble substructure, heat stick or vent pipe and combination of industrial thermal insulation material and heat stick should be adopted. If necessary, bridge should be used for substructure higher than 3.0 m.

2 Pebble substructure should be adopted for sections with groundwater developing, surface water developing or frozen soil swap. The thickness of the pebble is determined based on the ice content of the frozen soil, which should be 1.2–1.8 m. The pebble is separated as two layers, the top layer should be 0.4 m thick with type of 50–100 mm and the bottom layer may be 0.8–1.4 m with type of 150–200 mm. The compression strength of the rocks shall be higher than 30 MPa. Gravel layer should be installed below the pebble layer with a thickness of 0.3–0.5 m.

3 When heat stick is used in the substructure, the arrangement of the heat stick (single side or both sides) shall be determined based on the frozen soil condition and the highway alignment. The heat stick shall be placed at a distance of 0.10 m from the outside of the substructure with a longitudinal spacing which should be 1.5–2.5 times the equivalent radius of the heat stick which is determined based on the coagulation of the thermal energy. The heat stick should be buried at a depth of 1.0–2.5 m from the upper limit of the frozen soil. When the width of the substructure is less than 10 m, the heat stick should be placed vertically. When the width of the substructure is 10–12 m, the heat stick may be installed with an inclination angle

which shall be not greater than 15°.When the width of the substructure is greater than 12 m, the heat stick shall be installed incline or “L” shape heat stick shall be adopted.

4 When the local wind direction is perpendicular to the alignment, a vent pipe may be used. The material of the vent pipe should be prestressed steel-reinforced concrete with an inside diameter of 0.3–0.4 m. The spacing of the vent pipe shall be less than the cooling radius and two times of the outside diameter of the vent pipe. The buried depth of the vent pipe shall be greater than 3–5 times of the pipe diameter with a depth which should be 0.5–0.7 m from the ground surface. The length of the vent pipe outside from the side slope shall be greater than 0.3 m. The cooling radius need to be determined based on the calculation of the thermal energy.

5 Industrial insulation thermal material may be added in the substructure when the heat stick is not effective to control the thaw of the frozen soil.

7.12.6 The design of substructure in ice cone, frost heaving mound and thawing penetration lake-pond area shall meet the following requirements:

1 In a embankment below ice cone or frost heaving mound, a drainage ditch shall be built at its top to drain water from nearby ice cone or frost heaving mound. For a perennially thawing region, an insulation sewer shall be set to divert water away from the embankment when underground water flow is large. If necessary, a bridge need to be built.

2 The highway route on the section with ice cone or frost heaving mound, its plan shall be adopted with caution. When it must be adopted, drainage ditches, freezing ditches, stable seepage ditches and other facilities for cutting off groundwater shall be built outside the upper embankment slope foot no less than 20 m away. If ice amount is larger or there is an amount of underground water crossing through substructure and drainage is difficult, bridge should be built.

3 When substructure passes through a thawing penetration lake-pond, water may be pumped out and gravel or riprap may be used to reinforce silt if the area is not large. If necessary, a bridge need to be built.

4 For embankment in a swamp section, integrated measures of enhancing drainage, presetting settlement and eliminating freezing damage shall be adopted according to the principle of protecting perennially frozen soil based on swamp characteristics, water depth and perennially frozen soil type. If necessary, a bridge need to be built.

7.12.7 Roadbed drainage design shall meet the following requirements:

1 Surface water shall be drained out from the substructure in perennially frozen soil areas. A wide and shallow cross section shall be adopted for drainage ditch and intercepting ditch which should be far away from the toe of the substructure. If the drainage is difficult, a bridge shall be built.

2 For a frozen soil section with high ice content, a drainage ditch and intercepting ditch shall be avoided and water-stop ridge should be built. Measures of seepage prevention and insulation should be taken. If necessary, reinforcement measures shall also be taken.

3 In a rich ice or saturated ice frozen soil section, the distance from the inside of the drainage ditch, intercepting ditch or water-stop ridge to the slope toe of the insulation berm, cutting slope top or embankment slope toe (without insulation berm)

shall not be less than 5 m. In a section of ice layer that contains soil, the distance shall not be less than 10 m.

4 Freezing ditch, ice-collecting pit, ice-retaining dike, ice-retaining wall or sewer shall be built to drain underground water that is endangering the substructure according to underground water type, water amount and strata conditions. When the sewer is used to drain underground water, insulation measures shall be taken for both the sewer and inspection well.

7.12.8 Borrow pit and spoil bank design shall meet the following requirements:

1 The borrow pit (or field) shall meet the requirements of environmental protection in perennially frozen areas, be appropriately far away from highway and shall borrow earth intensively by sections.

2 The borrow pit (or field) shall be located in an area upside the embankment with less ice or in a sparsely vegetated ice frozen hillside or thawing area, bench and valley. The depth of the pit should not be greater than three fourth of the upper limit of the frozen soil. The site shall be flatten and the plant should be recovered.

3 The areas with saturated-ice and rich-ice frozen soil and ice layer containing soil shall not be used as borrow pit. The cutting soil with high ice content shall not be used as substructure or insulation berm fill.

4 The spoil bank shall be far away from the substructure. The field shall be flatten and the plant should be recovered. The rubbish with cement and asphalt shall be covered and that shall meet the requirements of environment protection.

7.12.9 The ground temperature and the settlement of the substructure shall be monitored for Class-II highway or above classes in perennially frozen soil area. The arrangement of the monitoring section shall be determined based on the landform, frozen soil condition, rock behavior and the substructure structure. For different landform sections, the minimum number of monitoring section should not be less than 2.If necessary, climate may also be monitored along the highway.

7.13 Substructure in Wind Drift Sand Area

7.13.1 Substructure in wind drift sand area shall be designed based on the following specifications:

1 For substructure design in a wind drift sand area, the local meteorology, topography and geomorphology, engineering geology and hydrogeology, highway building materials and ecological environment shall be investigated and collected, and the type of local desert and natural regions shall be ascertained.

2 Desert type, natural region and damage extent of the sand shall be combined to determine the highway route, substructure cross-sectional types and the sand protection system along the highway.

3 Substructure protection in an over-arid desert area with more serious wind sand flow shall meet the principles of obtaining materials from local resources, acting accordingly to local conditions and having comprehensive controls with the considerations of the geological condition along the highway, the climate condition

and the backfill material. The drift sand can be used to fill the substructure and also to protect the sand.

4 For the substructure in an arid flow desert area, side ditches and other drainage facilities may not be built. But in a section with the irrigation requirements, drainage design shall be considered and wide-shallow side ditches and large span bridges and culverts should be used.

7.13.2 Fill substructure shall be designed based on the following specifications:

1 Substructure in a wind sand drift area should mainly adopt low embankment. Generally, the embankment height should be 0.5–1.0 m higher than the average height of the sand dune within a range of 50 m on both sides of the embankment. When the route passes through sections with larger complex longitudinal dunes or composite dune chains, the fill height should be close to or slightly larger than the excavation height.

2 For the fill roadbed in a wind drift sand area, a gentle side slope or streamlined cross section should be adopted. A separated gentle slope should be used for Motorway and Class-1 highway while a convex median strip should not be adopted. The intersection of the shoulder and side slope should be made into a circular arc.

3 The embankment slope in a wind drift sand area shall generally be determined according to fill materials, fill height, wind direction, terrain and the highway and protection conditions according to Table 7.27. For Motorway and Class-1 highway in a slightly damp and semi-arid sand area, the embankment slope should be less than or equal to 1:3.

4 Strength and compaction degree of wind drift sand shall be in accordance with the requirements specified in Sects. 3.2 and 3.3. For pure Aeolian sand, geotextiles may be used to reinforce the substructure.

7.13.3 Excavation substructure shall be designed based on the following specifications:

1 Cutting with length longer than 200 m shall be avoided in wind drift sand area. For cutting in a wind drift sand area, an open and gentle slope cross section shall be adopted. When the alignment is perpendicular to the wind direction, the sand deposition stage should be widened.

Table 7.27 Side slope for filling substructure

Substructure height (m)		Side slope angle (1:n)	
		Motorway and class-1 highway	Class-2 and 3 highways
Flat site		1:3–1:6	1:3–1:6
Height of sand foundation	$h \leq 0.5$	1:3–1:6	1:3–1:6
	$0.5 < h \leq 2$	1:3–1:5	1:3–1:4
	$2 < h \leq 5$	1:2.5–1:4	1:2–1:3
	$5 < h$	1:2.5	1:2

2 The side slope of excavation shall be determined according to excavation depth, wind force and direction, terrain of highway sides and protective measures in accordance with Table 7.28.For Motorway and Class-1 highway in a slightly damp and semiarid sand area, the slope of cutting should be less than 1:3.

7.13.4 Cutting shall be protected according to the highway classification and road construction materials, along shoulders, side slope surfaces and areas in the range of 20–50 m outside of the slope top. The protection material may use the local material. If the climate is suitable, plant protection should be adopted. All protection structures shall be strong enough.

7.13.5 Design of borrow and spoil soil shall meet the following requirements:

1 Design of borrow and spoil soil shall protect the plant and crest at both sides of the highway. The borrow and spoil soil shall not be random. Borrow and spoil soil shall be performed in a main place when the sand with plant. For bare sand land, the sand dune and sand ridge may be used as borrow site; sand pit and sand depression can be used as spoil soil bank. The sand should not be excavated at flat sand. Protection shall be performed for borrow and spoil soil sites.

2 Soil should be borrowed at cutting sections or sand dune or sand ridge at the wind direction. When the sand is transported along the highway, the sand is obtained at the sides of the road, the borrow pit shall be 5 m away from the side slope toe at leeward side.

3 Excessive spoil soil shall be placed at the lower-lying area of leeward side and the distance to the top of the cutting shall not be less than 10 m.

7.13.6 The design of sand prevention engineering along sides of highway shall meet the following requirements:

1 Design of sand prevention shall use an overall plan with the usage of natural plants. The prevention measures shall be determined based on the local natural condition, the suitability of different type of prevention, local prevention experience to stop, fix and transport sand to combine the prevention of sand and plant. A comprehensive sand prevention system shall be established combined with the local sand prevention plan.

2 In a slightly damp and semiarid sand area, plant shall be used majorly to prevent sand with the supplement of engineering and chemical stabilization of sand. Plant including tree, bush and grass should be used for sand protection.

Table 7.28 Side slope for cutting substructure

Cutting depth (m)	Side slope angle (1:n)	
	Motorway and class-1 highway	Class-2 and 3 highways
$h \leq 0.5$	1:4–1:8	1:3–1:8
$0.5 < h \leq 2$	1:4–1:6	1:3–1:5
$2 < h \leq 5$	1:4–1:5	1:3–1:4
$5 < h$	1:4	1:3

3 In dry and desert area, a combination of engineering sand prevention, chemical prevention and plant prevention should be adopted. The plant prevention first and then engineering prevention, the plant can use bush and subshrub.

4 In extremely dry desert area, a comprehensive prevention system shall be established at the substructure and both sides for sections with sand flow or source rich sand flow to stop, block and transport sand. For sections with fixed sand dune and sand flow, the sand transport should be a major method combined with local treatment of sand dune. For other areas, the prevention system shall be installed based on the sand flow intensity and the damage of sand.

5 For dry, extremely dry desert areas, a plant prevention method may be adopted to build sand prevention barrier when the irrigation is available.

7.13.7 Sand prevention design shall meet the following requirements:

1 One to two sand barrier should be constructed at 80 and 150 m from the substructure in the wind direction for the sand prevention design. The sand barrier may use fence shape, wall shape, embankment shape and belt shape with 1.2–1.5 m out from the ground surface. The sand barrier may use the local material. If applicable, plant may use trees and bushes.

2 Sand stabilization design should use plant and engineering methods. The width of the sand stabilization belt shall be determined based on the sand source, sand flow intensity and sand dune movement behavior. At the wind direction, the stabilization belt should be 60–200 m in width, may be 0 m for single direction wind at leeward wind side and shall not be less than 50 m at reverse direction of leeward wind side. Sand stabilization using engineering method may use the local material to form grid and belt shape barrier or tile sand.

3 In a section within Gobi desert, deposit soil or saline soil areas, the ground surface shall be kept flat at a range of 20–30 m on both sides of the substructure and the gentle slope cross section or other sand transporting method should be used. The ground protrusions shall be eradicated and flattened to form a flat zone.

4 The plant prevention belt should use grass, trees and bushes. The trees should use both from early and later stage with local tress. The width of plant prevention belt may be determined based on the wind drift sand flow intensity.

5 When conditions permit, a vegetation protective belt should be set up outside the shelter belts on both sides according to the severity of wind and sand. The width of the vegetation protective belt shall generally be no less than 300 m at the windward side of the substructure and no less than 100 m at the leeward side.

6 When plant prevention sand measures are to be used, a suitable plant species shall be chosen that combines with local site conditions to determine appropriate plant structures and planting patterns. In the meantime, sound irrigation measures and management system shall be established.

7. The sand prevention engineering system shall set 2–4 fire barriers with a width of 2–3 m per kilometer along both sides of the highway.

7.14 Substructure in Snow Disaster Area

7.14.1 For substructure design in a snow disaster area, the following material shall be collected. The causes of snow hazards shall be analyzed, the types of snow hazards shall be determined and the appropriate treatment measures shall be provided.

1 Accumulated ice area and the landform, plant and climate during the snow flow.

2 Wind power, wind direction, wind speed gradation and frequency and lasting time, quantity of snow, depth of snow and density of snow.

3 Distribution, crack location and frequency of sand slide. If necessary, the snow accumulated area map and the snow slide route map in longitudinal direction shall be drawn.

7.14.2 The alignment shall avoid the snow disaster area and use sunny slope in hilly area. If the snow disaster area could not be avoided, a shortest alignment shall be adopted at snow light place. The alignment should be parallel or at most of 30° from the snow flow direction and the protection should also be adopted.

7.14.3 Snow prevention design shall use the principle of prevention first and the combination of prevention and improvement. A comprehensive method shall be adopted for preventing, stabilizing, blocking, directing and removing snow using plant or engineering measures.

7.14.4 Design of substructure in snow disaster area shall be based on the local wind snow and landform condition, and a suitable cross section should be adopted. Fill is good and excavation is not applicable for snow disaster area. Motorway and Class-1 highway should choose the integrated substructure or the separated gentle slope substructure which is conducive to the movement of snow flow. Various facilities should not be set up in the substructure and the road shoulder, and the obstacles and structures on both sides of the substructure within 20 m from the foot of the slope shall be removed.

7.14.5 The minimum embankment height in snow drift area shall not be less than the thickness of snow in a frequency of 50 years plus the safety height which may be shown in Table 7.29.The slope angle at windward side should be 1:3–1:4, while it should be gentled to 1:4 when the single directional wind is strong.

Table 7.29 Safety substructure height at different extent of snow disaster

Extent of snow disaster	Safety height (m)	
	Motorway and Class-1 highway	Class-2 highway and below
Light	0.5–0.7	0.5–0.7
Medium	1.0–1.5	0.7–1.0
Heavy	1.5–2.0	1.0–1.5

Note 1. Snow disaster is classified as weak, medium and heavy based on the snow quantity, drift snow, wind direction to the highway and the maximum depth of the accumulated snow

2. The value of safety height shall be larger if the snow accumulation period is long. If the line position is in low terrain, the high value is suitable. If the line position is in high terrain, the low value is suitable

7.14.6 When excavation must be used for drift snow area, deep excavating shall be avoided. A gentle slope should be adopted for excavation sections and the slope angle should be 1:2–1:5.When the cutting slope is higher than 2.0 m, a snow accumulation stage should be constructed at both sides of the side slope with a minimum width of 5 m and the stage top surface shall be at least 1.5 m higher than the surface of the highway. When the alignment is 45° to 90° to the wind direction, an open cutting substructure shall be used with gentle slopes and the snow accumulation stage should also be constructed.

7.14.7 At drift snow sections with half fill and half excavating, the alignment shall be an acute angle with the wind direction and the excavation width should also be increased at least to 2 m. When the alignment is 45° to 90° to the wind direction, an open cutting substructure shall be used with gentle slopes.

7.14.8 Deign of prevention and control measures of snow drift shall meet the following requirements:

1 Snow protection plantation belt may be planted at one side or both sides of the substructure when conditions are appropriate. The width of the belt should not be less than 50 m. The tree-shrub mixed types should be adopted. Several belts should be adopted and the space between belts should be 20–50 m and the width of a single belt should be 20 m. The clear distance from the snow protection plantation to the slope toe of the substructure may be set as 10 times the snow protection plantation height, but it shall not be less than 25 m.

2. Snow Prevention fence may be classified as fixed and mobile fence, and the design shall follow the specifications including:

(1) For sections where the amount of snow is small but duration is long and wind direction changes little, a fixed snow prevention fence may be adopted. The height of the fixed-type prevention fence shall be determined according to wind force and snow volume, but should not be less than 3 m. The distance from the substructure edge to the prevention fence shall be determined according to the length of the snow pack dike behind the fence, which should be 30–50 m.
(2) For sections in which wind direction varies and the wind force and snow amount are large, a movable prevention fence may be adopted. The height of the movable prevention fence should be 1–2 m. The initial location of the fence should be 20–50 m away from the substructure edge.
(3) The snow prevention fence shall be built at the upwind side and shall be perpendicular to the winter dominant wind direction. When the terrain is open and snow amount is extremely large, two rows of fences may be built and the space between them should be 50–80 m.

3. Wind baffler may be classified as downward wind baffler and lateral wind baffler, and the design shall follow the specifications including:

(1) Downward wind baffler may be suitable for sections in which the intersecting angle of highway with the dominant wind direction is larger than 30° and the slope grade of the upwind hillside is less than 40°.Otherwise,the lateral wind baffler should be adopted.

(2) The location of the wind baffler shall be determined according to the local dominant wind direction, cross section type of the substructure and terrain conditions. The downward wind baffler should be laid near the shoulder of the upwind side with a distance of 0.75 m to the outside of the shoulder. The substructure side slop at windward side should be flat. The lateral wind baffler should be laid at outside of the substructure edge with the upwind side no less than 15 m.

4 For sections where snow pack is less and a prevention fence should not be set, a snow retaining wall or prevention dike may be built at the upwind side. The height of the prevention dike (wall) may be determined according to the snowfall amount.

5 An open cut tunnel should be constructed at sections with high wind speed, low visibility and the angle of alignment to the wind direction greater than 60°.Drift snow disaster prevention, ventilate and fire protection should be designed for entrance and exit.

7.14.9 The fill substructure should be used in the snow drift area, and the embankment height should be greater than the maximum snow accumulation thickness when snow drift occurs. In accordance with the principles of stabilizing snow cover on slopes, changing the direction of avalanche movement, slowing down avalanche movement and removing snow cover, facilities such as horizontal steps, snow stabilization fences, mounds, snow guide dikes (walls), snow retaining walls and snow forest belts should be set up. In more serious snow drift areas, snow-proof corridors, open cuts, tunnels and other shelter structures can be used.

7.14.10 Design of snow slide prevention shall follow the specifications including:

1 Horizontal steps may be used to prevent the small snow slide with a route of less than 45° to the ground surface transversal slope, no landslide or debris flow. The step width shall be determined based on the maximum depth of the accumulated snow and the slope angle.

2 Snow stable fence may be used to stable the snow in snow slide areas with steep slopes and difficulty to excavate horizontal steps. Snow stable fence should be installed in multiple rows, and the highest row should be located at the place close to snow crack point and below the snow edge. The fence height shall be 0.5 m higher than the maximum accumulated snow thickness.

3 Snow prevention fence may be used to prevent snow slide in snow slide areas with gentle slopes. Mutiple rows of snow prevention fence shall be necessary whose strength shall be determined based on the calculation of the impact force. The fence shall be 1.0 m higher than the middle part of the snow during sliding.

4 Snow protection plantation belt may be used from the source of snow slide to the active zone of snow slide. Proper tress need to be planted from top down in periods, while at early stage, engineering measures may be used to combine with the Snow protection plantation belt.

5 Earth mound may be used to decrease the thickness of the soil layer and the speed of snow sliding in sliding ditches with slope angle less than 30°.Earth mound should be located at the turning point of the longitudinal slope on the snow slide route which shall be at least 1.0 m higher than the front of the snow slide.

6 Snow directing embankment may be placed in a wide snow slide trench. The angle of snow direction embankment with the snow slide shall be less than 30°.The height shall be 0.5–1.0 m higher than the front of the snow sliding.

7 Snow retaining wall shall be constructed in the active and accumulated zone of the snow slide. Snow retaining wall may be constructed using laid dry (laid wet) pebbles or steel-reinforced concrete and the strength of the wall shall be calculated based on the impact force of the snow slide. The retaining wall in the active zone shall be 0.5 m higher than the front of the snow slide and the maximum thickness of the snow in the accumulated zone.

8 Snow prevention corridor may be used for sections with severe snow slide. The top of the snow prevention corridor shall meet the requirement of the impact force induced by snow sliding. The clearance of the corridor shall meet the requirements of tunnel clearance in the current *Highway Engineering Technical Standards* (JTG B01).

7.15 Substructure in Ice Drift Section

7.15.1 Design of substructure in ice drift section shall meet the following requirements:

1 For substructure in the ice drift section, the local terrain, geology, meteorology, water source of the ice drift, type and scale, hazard and experience in prevention and control shall be investigated. The main contents include the terrain and type of water source, flow volume, soil type and thickness, freezing and thawing cycles and depth, type and scale and damage potential of ice drift and its relationship with routes that can be selected.

2 Substructure design in the ice drift section shall combine prevention and control measures, with prevention as the top priority to avoid ice drift and influence the safety of the highway. The type of the selected facilities shall be convenient for maintenance.

3 In frozen ice drift or alpine regions, the substructure should be built on the dry sunny slopes as much as possible and an embankment or shallow cutting should be adopted.

4 For an ice drift in a river valley, the substructure shall be raised and a large span bridge shall be built to avoid ice drift flowing on the pavement.

5 For an ice drift on a hillside, diverting the hillside water away from the substructure, widening and deepening the side ditch at the upper slope and building facilities such as ice-retaining wall, ice-collecting pit, ice retaining bank and ice-collecting ditch may be adopted. A sewer, blind ditch and other underground drainages shall be built when the underground water on the hillside is larger. A culvert with large clearance may be built at the ice-collecting ditch or ice-collecting pit in order to drain away melting ice water.

6 Substructure engineering shall avoid interference with the existing natural drainages and should not cut aquifer. When drainage, retaining, interception and other control measures are adopted, the natural drainage system shall be kept unimpeded.

7.15.2 When the assembled ice amount of the ice drift is not large, a raised substructure may be used. The height shall be larger than the maximum height of obstructed ice plus 0.5 m, while construction materials may be selected from crushed gravel with good water stability.

7.15.3 When a bridge is adopted to cross the ice drift, the bridge clearance shall be the highest level of ice drift on record plus the back ice height and safety height of 0.5 m. When the clearance for the bridge is limited, ice directing embankment, river trench or ice-collecting pit may be adopted.

7.15.4 For the ice drift on the alluvial fan or flatter slope, an ice-collecting ditch may be built on the outside of the upper slope of the substructure. The cross-section of the ice-collecting ditch shall be determined according to the terrain, geology, water volume and ice accretion amount, the depth of the ditch and the width of the trench bottom should be 0.8–1.2 m. The connections between drainage facilities should be smooth. The ice-retaining dike should be 0.8–1.2 m high and the width of the dike top should be 0.6–1.0 m. The side slope should not be steeper than 1:1.5.When the dike is constructed using laid dry pebble, the slop could be steeped to 1:0.5.

7.15.5 Ice-retaining wall shall be located on the outside of the side ditch and built with mortar flagstones or stone blocks with height of 1–2 m. When the amount of ice accretion is large, the ice-collecting pit may be built on the outside of the ice-retaining wall. The width of the collecting pit should be 1.5–3.0 m. For an ice-collecting pit at the soil section, a dry-laid rubble low wall may be built at the foot of the side slope based on the permeability and soil conditions of the slope surface. The side ditches shall be protected by mortar flagstones.

7.15.6 When underground water is exposed, sewer, blind ditch and other underground drainages may be used to divert underground water away from the substructure. Underground drainages shall be located below the frozen depth and the exit shall be at least 0.5 m higher than the ground surface. The outlet shall be kept warm or a drainage ditch with a longitudinal slope greater than 10% shall be excavated.

7.16 Substructure in Mined-Out Area

7.16.1 Design of substructure in mined-out area shall follow the specifications including:

1 For substructure design in a mined-out area, the nature environment, distribution and exploitation of mineral resources, ground deformation and movement and other information along the highway shall be collected and investigated. The distribution, size and change characteristics of the mined-out area and physical and mechanical properties of rock and soil mass in the relevant strata shall be ascertained through investigation, mapping, integrated exploration, and ground deformation monitoring.

2 The highway shall avoid widely distributed, large scale and difficultly treated mined-out areas. If it is difficult to avoid, substructure should be adopted and the connection of substructure with the bridge shall avoid the large settlement area.

3 For substructure design in mined-out area, design options shall be compared by combining local environment feature, engineering geological conditions, materials distribution and supply, resource planning, schedule requirements and other factors and based on the distribution and deformation in the mined-out area. The substructure and pavement shall be designed well.

4 Design of substructure in mined-out area shall consider the influence of the vehicle loading, the weight of substructure and pavement and the combination of the ground surface settlement and the substructure settlement and use the substructure structures including light weight embankment and geosynthetic-reinforce embankment. The treatment of the mined-out area shall be adopted to meet the requirement of the substructure construction and to ensure the stability of the substructure.

7.16.2 Ground surface evaluation shall be carried out for the design of substructure in mined-out area with the principle of the combination of qualitative and quantitative. Basis of the evaluation includes type of mined out area, size, characteristics of overlying rocks, ratio of thickness and depth of mining, inclination angle of coal layer, mining time and hydrological and geological conditions. The remained ground deformation shall be predicted based on the combination of the mining condition method, the ground deformation prediction method and ground deformation monitoring method to evaluate the stability of the mined-out area.

7.16.3 Standard for the stability evaluation of the mined-out area shall meet the following requirements:

1 The ground surface deformation in the mined-out area shall meet the requirements in Table 7.30.When the ground surface deformation could not meet the requirements, the mined-out area shall be treated.

2 Sections with ground surface inclination in mined-out area greater than 10 mm/m, ground surface curvature greater than 6 mm/m^2 and ground surface lateral deformation greater than 6 mm/m should not be used as substructure construction site.

7.16.4 When highway is overlying the mines, the mines shall design follow the specifications including:

1 In the coal mines which have not been mined, protection coal piles shall be constructed for Motorway, Class-I highway, tunnels, extra-long span bridges,

Table 7.30 Allowable ground surface deformation in mined-out areas

Highway class	Ground surface inclination (mm/m)	Lateral deformation (mm/m)	Ground surface curvature (mm/m^2)
Motorway and Class-I highway	≤3.0	≤2.0	≤0.2
Class-II highway and class below	≤6.0	≤4.0	≤0.3

long and middle span bridges and sections can be affected by mining with serious landslides.

2 Protection coal piles shall be installed outside of the enclosure zone, and the width shall be designed to meet requirements as follows:

(1) For embankment, the enclosure zone should reach 1 m from the embankment side slope toe at both sides, while that for cutting is at the cutting slope top at both sides;
(2) The enclosure zone is along the highway at both sides. The width of the enclosure zone for Motorway is 20 m and 15 m for Class-1 highway.

3 The outside boundary of the coal protection piles for inclined coal layers is determined using vertical section method, normal line method or digital projection with heights method based on the upward and downward moving angle and the moving angle in the loose layer.

7.16.5 Treatment of mined-out areas shall follow the specifications including:

1 Treatment length shall be the actual distribution length of the mined-out area along the highway axis. The treatment width shall be the substructure width or the structure width and the treatment depth should reach the weathered rock.

2 Treatment area for other types of mined-out area shall follow the requirements including:

(1) When the mined-out areas are thick, the treatment length shall increase the influence width of the movement angle of the overlying rocks. The treatment length along the highway may be calculated using Eq. (7.16.5.1):

$$L = L_0 + 2h\cot\alpha + H_U\cot\beta + H_L\cot\gamma \qquad (7.16.5.1)$$

where:

L—the width in the horizontal direction perpendicular to the highway axis (m);

L_0—the length of mined-out areas along the middle line of highway;

H_U—the thickness of the overlying rockbed of the mined-out area in uphill direction (m);

H_L—the thickness of the overlying rockbed of the mined-out area in downhill direction (m);

α—the movement angle of the overlying rockbed of the mined-out area in the loose layer (°);

β—the movement angle of the overlying rockbed of the mined-out area in the uphill direction (°);

γ—the movement angle of the overlying rockbed of the mined-out area in the downhill direction (°).

(2) The treatment width includes the width of the substructure bottom, the enclosure zone width and the influence width of rocks in the mined-out area. For lateral layer, the width may be calculated using Eq. (7.16.5.2).When the inclined rock layer is perpendicular to the rock layer, the width at any point along the highway may also be calculated using Eq. (7.16.5.2).When the inclined rock layer is

parallel to the rock layer, the width at any point along the highway may be calculated using Eq. (7.16.5.3).When the inclined rock layer is skew to the rock layer, the width at any point along the highway may be calculated using Eq. (7.16.5.4).

$$B = D + 2d + 2(h\cot\alpha + H\cot\delta) \tag{7.16.5.2}$$

$$B = D + 2d + 2h\cot\alpha + H_{\mathrm{U}}\cot\beta + H_{\mathrm{L}}\cot\gamma \tag{7.16.5.3}$$

$$B = D + 2d + 2h\cot\alpha + H_{\mathrm{U}}\cot\beta' + H_{\mathrm{L}}\cot\gamma' \tag{7.16.5.4}$$

$$\cot\beta' = \sqrt{\cot^2\beta\cos^2\theta + \cot^2\delta\sin^2\theta} \tag{7.16.5.5}$$

$$\cot\gamma' = \sqrt{\cot^2\gamma\cos^2\theta + \cot^2\delta\sin^2\theta} \tag{7.16.5.6}$$

where:

B—width of substructure along the highway (m);
D—width of the substructure bottom (m);
d—the enclosure zone width (m),generally 10 m;
H—thickness of the rock layer (m);
h—thickness of the loose layer (m);
δ—the movement angle of the overlying rockbed of the mined-out area in the strike direction (°);
β'—the movement angle of the overlying rockbed of the mined-out area in the uphill direction (°);
γ'—the movement angle of the overlying rockbed of the mined-out area in the downhill direction (°);
θ—acute angle of the boundaries of the enclosure zone with the inclined line of the mine layer (°).

(3) Treatment depth is from the ground surface to the floor of the mined-out area (or coalbed) and shall not be less than 3 m when the treatment is inside of the mined-out area; when the treatment is at outside of the mined-out area to the influence area of the rock layer movement, the treatment depth shall be calculated using Eq. (7.16.5.7):

$$h_{\mathrm{t}} = H - l\tan\delta_{\mathrm{O}} + h' \tag{7.16.5.7}$$

where:

h_{t}—treatment depth outside of the mined-out area (m);
H—depth of the mined-out area, namely the thickness of the overlying rock layer (m);
l—distance from the grouting hole to the boundary of the mined-out area (m);

h'—treatment depth below the crack, should use 20 m;

δ_O—movement angle of the overlying rock layer outside of the mined-out area (°).

7.16.6 For the highway design of the mined-out area, treatment plans shall be selected based on the formation time of the mined-out area, depth, mined thickness, coal-mining method, roof lithology and mechanical properties, hydrogeology, engineering geological conditions and so forth. The treatment plan shall follow the specifications including:

1. The shallow mined-out area and the cutting slope should be treated with backfill.

2 Dry-laid (mortar) flagstone may be applied to the treatment of the mined-out area under conditions of non-complete collapse, larger space, small depth, good ventilation, having manual operation and material transportation. The substructure of the general section may be backfilled with dry-laid flagstone whose compressive strength shall be no less than 10 MPa. The road section with structures shall be constructed with mortar flagstone with compressive strength shall be no less than 15 MPa.

3 The dynamic compaction may be used for mined-out areas with depth less than 10 m and perfect overlying rock layer with low strength.

4 Backfilled with pebble, supporting and grouting may be used to treat mined-out areas with great depth and tunnels.

5 Bridge crossing may be adopted for the mined-out area where the coal-mining scale is small and the mining depth is less than 100 m.

6 Full-filled grouting method should be adopted in a mined-out area with a larger coal-mining scale and mining (buried) depth of less than 250 m. For a mined-out area whose buried depth is more than 250 m, deciding whether to adopt full-filled grouting method should be based on exploitation features, hydrogeology, engineering geology and its damage degree to highway engineering.

7.16.7 Motorway and Class-I highway shall be monitored during construction, and the monitoring design shall use the following specifications:

1 The design of monitoring the deformation system and the monitoring locations shall be based on the characteristics of the mined-out area and the overlying rock layer combing with the type of highway. The monitoring should include lateral movement monitoring, vertical movement monitoring, structures inclination monitoring and cracks monitoring.

2 The Monitoring period shall be started at the site investigation stage and finished at least one year after the completion of the construction.

7.17 Substructure in Coastal Area

7.17.1 Substructure in coastal area shall meet the following requirements:

1 Design of substructure in a coastal area shall be based on geographical environments and features of the substructure location. The topography, geomorphology, geology, hydrology, meteorology and other factors shall be considered together with

construction conditions and materials supply. The design elevation of the substructure shall be reasonably determined and the type of cross-section and protection work shall be properly selected to ensure the integral stability of the substructure.

2 Road in coastal area shall constructed at sections with narrowest surface, low water, small wave, flat beach and good geological conditions.

3 Water passing structures should be provided when a large water head difference exists on both sides of the embankment. When piping or underground erosion may occur, drainage facilities shall be provided at the underpart of the slope located at the lower water level side. The slope shall be flattened or berms and a cutoff wall shall be provided along the inside of the embankment.

4 Embankment fill shall consist of materials with good permeability including coarse-grained or huge grained soil. If there is frost heave, the water retaining wall shall be installed at the outside of the submerged embankment slope toe.

5 When the substructure in coastal area is over soft soil, ground improvement shall be needed which shall follow the specifications in Sect. 7.7.

7.17.2 Design elevation of coastal substructure shall meet the following requirements:

1 The design elevation of a coastal substructure shall not be less than the design tide water level in the peak frequency plus the wave attack height plus 0.5 m safety height. The design frequencies of high tide water level for classified highways are specified in Table 7.31.When the specified requirements cannot be met, a wave wall shall be provided.

2. The design wave criteria shall meet the following requirements:

(1) The design wave return period standard: 50 years return period shall be adopted for the Motorway,Class-1 and 2 highways;25 years return period shall be adopted for Class-3 and 4 highways.
(2) When calculating the strength and stability of the bracing structure and slope protection engineering of the coastal substructure, the wave trains cumulative frequency of design wave height should be determined according to Table 7.32.

7.17.3 The cross section of the coastal substructure shall be determined based on the water depth, wave height, foundation condition, backfill material behavior, construction condition, and usage requirement. Generally, sloping type should be adopted with exception of limitation of backfill material and other requirements for usage using straight wall type.

Table 7.31 Substructure design high tide water level frequency

Highway classification	Motorway	Class-1 highway	Class-2 highway	Class-3 highway	Class-4 highway
Substructure design high tide water level frequency	1/100	1/100	1/50	1/25	Determined by the specific conditions

Table 7.32 Wave train cumulative frequency standard

Coastal embankment type	Parts	Calculating contents	Wave train cumulative frequency $F(\%)$
Sloping	Parapet walls, blocks at the embankment top	Strength and stability	1
	Slope protection block stone, slope protection block mass	Stability	13
	bottom protection block stones	Stability	13
Straight Wall	Superstructure, wall and pile	Strength and stability	1
	Foundation bed, bottom protection block stones	Stability	5

Note When calculating the stability of embankment with inclined slope reinforced by the pebble, the average ratio of wave height to the water depth is less than 0.3 and the wave train cumulative frequency should use 5%

7.17.4 The coastal embankment slope rate shall be determined based on the properties of fill, embankment height, immersion depth, protection type and marine hydrological conditions. The slope rate should not be larger than 1:1.75.

7.17.5 Coastal embankment slope protection shall meet the following requirements:

1 The materials of slope protection shall adopt lath stones, block stones, concrete profiled blocks and so forth based on the water depth, wave characteristics, construction conditions and materials. To weaken the destructive effects of waves on the embankment and improve the stability of the embankment slope, measures such as the breakwater prismoid, longitudinal dike and submerged dike in front of the embankment shall be taken. All protection engineering measures shall be resistant to marine and geological erosion and to frost and striking of ice-jam flood in cold regions.

2 The rock mass of slope protection cushion may be 1/10–1/20 of the mass of rocks for slope protection and also meet the requirement of slope stability under wave acting during construction. The thickness of the cushion should not be less than 400 mm.

3 A rock mound shall be provided at the bottom of the slope protection along the open sea side. Its top surface elevation shall be higher than the construction water level and its top surface width should not be less than 1.0 m.

4 Proper measures of protecting the slope bottom along the open sea side shall be taken based on maximum scour depth, terrain and foundation type. The stone thickness for the bottom protection shall not be less than 1.0 m and the width shall not be less than 5.0 m. The thickness of gravel cushion should be placed underneath

the protection rock at the sandy seabed should be not less than 0.3 m, and the width of the protection rock shall be determined based on the erosion.

7.18 Substructure in Reservoir Area

7.18.1 The design of substructure in reservoir areas shall follow the specifications including:

1 The design information of the reservoir water level and meteorological data of the reservoir area shall be collected. The terrain and topographic features of reservoir banks, geological structure, stratum lithology of the reservoir bank composition and changes of the groundwater level shall be examined. The stability of slopes in the valley shall be ascertained. It shall be verified whether adverse geologic phenomena such as landslide or collapse exist. The impact of reservoirs on the slope stability shall be analyzed and evaluated.

2 The design of substructure shall consider the changes of properties of soils, which is induced by the reservoir water soakage, permeation, water level fluctuation, wave attacking, water erosion, reservoir banks collapse, siltation and rise of groundwater caused by stasis, etc. The stability of the substructure and reservoir should be evaluated with reinforcement measures to ensure the stability.

7.18.2 Types of substructure cross-sections and requirements of fill shall meet the following requirements:

1 Embankments shall be designed according to requirements of immersed substructure. The rate of the side slope below the design water level should not be larger than 1:1.75.When the height of the slope is large, the step-shaped cross section should be adopted.

2 Permeable materials with small compression deformation and good water stability shall be adopted to fill the substructure and the compaction degree of backfill material shall be strictly controlled. If permeable materials are lacking, the parts of embankments immersed in the water of the reservoir shall be filled by permeable materials while the portion above the water level may be filled by fine-grained soils.

3 When the substructure is submerged in a long time and high water head difference exists at two sides, the submerged portion should be filled with difficultly-weathered rocks, side slope at lower water head should be flatten and berm should be widened.

4 When the seepage velocity and seepage pressure are higher and erosion might occur, side slopes should be flattened and the drainage facilities should be provide at the side with low water level.

7.18.3 Stability analyses of embankments and reservoir banks shall meet the following requirements:

1 To analyze the embankment stability, the seepage pressure and erosion effect on the embankment slope caused by the stable seepage that is induced by the head difference between upstream and downstream and the transient seepage that is induced in sharp fall in water level shall be considered. Soil embankment shall be checked for

the worst situation of seepage in the embankments. If necessary, flow nets shall be applied to the calculations.

2 The strength parameters of soil shall be determined by quick shear test for soil above the ground water and saturated quick shear test for soil below the groundwater respectively after the soil has been rammed (above the infiltration curve adds to the rising height of groundwater caused by stasis).The physical parameters shall be determined for above and below the groundwater level conditions separately.

3. Ice loads shall be considered in an ice frozen area and ice drift area. At the upstream sites of reservoirs, the water erosion shall be considered with the high flow velocity.

4 The stability safety factor of the slope shall not be less than the values in Tables 3.11 and 3.15.The factor of safety shall not be less than 1.05 when the water level fluctuations and seismic effects are simultaneously considered.

7.18.4 The type of substructure slope protection shall be determined in accordance with the requirements such as reservoir type, wave force, and substructure location. Side slopes may be protected using laid dry (wet) pebble, concrete and gabion. Impermeable and inverted filter layer shall be established. When soakage and other factors affect stability, the retaining walls, longitudinal or spur dick with erosion prevention and secondary levees may be used to reinforce the substructure. Various protection works shall be in harmony with the surrounding landscape.

7.18.5 Ground improvement of the substructure shall follow the specifications including:

1 The foundations including soft soil foundation in submerged substructure, weak soil due to soak of reservoir water and groundwater, collapsible loess foundation, liquefiable silt and fine sand under dynamic loading shall be reinforced.

2. When the seepage at the substructure bottom affects the stability, cover layer at side slope toe, seepage prevention wall and curtain shall be adopted to prevent seepage.

7.18.6 When the stability of the substructure is endangered by the reservoir bank collapse, appropriate protective measures shall be taken based on highway location, bank soil, bank height and slope ratio, immersion depth, reservoir sedimentation and so forth. The type of protection facilities against bank collapse shall be selected reasonably according to the destructive effects of waves and topographical and geological conditions.

7.18.7 The protective length shall be determined according to how the reservoir bank slope is affected by waves. The protection work shall provide appropriate safety distance at both ends and shall be embedded into the reservoir bank or substructure slope.

7.19 Substructure in Seasonally Frozen Soil Area

7.19.1 Design of substructure in seasonally frozen soil area shall follow the specifications including:

1 The investigation shall be conducted to collect the average annual air and ground temperature, frozen index, standard frozen depth, the damage situation of the local substructures due to frozen soil and improvement experience. The frost heave class and its damage to the substructure shall be evaluated based on the understanding of the distribution of the seasonally frozen soil layers, soil properties, groundwater level and the frozen water capillary height.

2. The substructure in seasonally frozen soil area should be filled and not be excavated. The alignment should be located at sunny slopes.

3 The substructure height shall be determined based on the effects of climate, landform, geological and drainage conditions, backfill material on the frozen damage of the substructure and the pavement.

7.19.2 Classification of frost heave of the seasonally frozen soil shall meet the following requirements:

1 The frost heave class shall be determined based on the average frost heave degree using Table 7.33.

2 The average frost degree is calculated using Eq. (7.19.2):

$$\eta = \frac{z}{H_d} \times 100(\%) \tag{7.19.2}$$

where:

z—frost heave (mm);

H_d—frozen depth (mm) without frost heave.

3. The frost heave behavior of the seasonally frozen soil shall be classified as Table 7.34.

7.19.3 The frost heave of the substructure standard shall meet the following requirements:

1. The frost heave of the substructure shall be calculated using Eq. (7.19.3) and the calculated depth shall be the frozen depth.

$$z_j = \sum_{i=1}^{n} h_i \eta_j \tag{7.19.3}$$

where:

z_j—frost heave (mm);

Table 7.33 Frost heave class of the seasonally frozen soil

Average frost heave degree η (%)	Frost heave class	Frost heave type
$\eta \le 1$	I	No frost heave
$1 < \eta \le 3.5$	II	Weak frost heave
$3.5 < \eta \le 6$	III	Frost heave
$6 < \eta \le 12$	IV	Strong frost heave
$\eta > 12$	V	Extremely strong frost heave

Table 7.34 Frost heave class of the seasonally frozen and thaw soil

Soil type	Natural water content before frozen W (%)	Minimum distance of the groundwater level to the design frozen depth before frozen h_w (m)	Average frost degree η(%)	Frost heave class	Frost heave type
Gravel, crushed stone, coarse sand, medium sand [the content of fines (particle size less than 0.075 mm) is not greater than 15%], and fine sand (content of fines is not greater than 10%)	Unsaturated	NA	$\eta \le 1$	I	No frost heave
	Saturated	Without water separation layer	$1 < \eta \le 3.5$	II	Weak frost heave
	Saturated	With water separation layer	$3.5 < \eta$	III	Frost heave
	$W \le 12$	>1.0	$\eta \le 1$	I	No frost heave
	$12 < W \le 18$	≤1.0	$1 < \eta \le 3.5$	II	Weak frost heave
		>1.0			
	$W > 18$	≤1.0	$3.5 < \eta \le 6$	III	Frost heave
		>0.5			
		≤0.5	$6 < \eta \le 12$	IV	Strong frost heave
Silty sand	$W \le 14$	>1.0	$\eta \le 1$	I	No frost heave
		≤1.0	$1 < \eta \le 3.5$	II	Weak frost heave
	$14 < W \le 19$	>1.0			
		≤1.0	$3.5 < \eta \le 6$	III	Frost heave
	$19 < W \le 23$	>1.0			
		≤1.0	$6 < \eta \le 12$	IV	Strong frost heave
	$W > 23$	NA	$\eta > 12$	V	Extremely strong frost heave
Silt	$W \le 19$	>1.5	$\eta \le 1$	I	No frost heave
		≤1.5	$1 < \eta \le 3.5$	II	Weak frost heave
	$19 < W \le 22$	>1.5			
		≤1.5	$3.5 < \eta \le 6$	III	Frost heave
	$22 < W \le 26$	>1.5			
		≤1.5	$6 < \eta \le 12$	IV	Strong frost heave
	$26 < W \le 30$	>1.5			

(continued)

Table 7.34 (continued)

Soil type	Natural water content before frozen W (%)	Minimum distance of the groundwater level to the design frozen depth before frozen h_w (m)	Average frost degree η(%)	Frost heave class	Frost heave type
		≤1.5	$\eta > 12$	V	Extremely strong frost heave
	$W > 30$	NA			
Clay	$W \le W_p + 2$	>2.0	$\eta \le 1$	I	No frost heave
		≤2.0	$1 < \eta \le 3.5$	II	Weak frost heave
	$W_p + 2 < W \le W_p + 5$	>2.0			
		≤2.0	$3.5 < \eta \le 6$	III	Frost heave
	$W_p + 5 < W \le W_p + 9$	≤2.0			
		>2.0	$6 < \eta \le 12$	IV	Strong frost heave
	$W_p + 9 < W \le W_p + 15$	>2.0			
		≤2.0	$\eta > 12$	V	Extremely strong frost heave

Note 1. $W_p$is the soil plastic limit (%); W is the average natural water content in the frozen layer before frozen
2. Frozen saline soil is not included in this table
3. When the plasticity index is greater than 22, the frost heave class should low one level
4. When the content of particles with size smaller than 0.005 mm is higher than 60%, the soil is non-frost heave soil
5. When the gravel as fill material and its mass is greater than 40% of the total mass, the frost heave is classified based on the filled material
6. Water separation layer is referred to the water separation layers at and above the bottom of the seasonally frozen soil layer

h_i—soil thickness of layer i in the frozen zone(mm);
η_i—frost heave degree in layer i (%);
n—number of soil layers.

2. The Frost Heave of the Substructure Shall Meet the Requirements in Table 7.35.

7.19.4 The design of backfill for the substructure shall be in accordance with the substructure height, surface water level, groundwater level, total allowable frost heave and pavement type as shown in Table 7.36 to determine the backfill material. The preferable material with good frozen resistance including medium coarse sand, crushed stone, sandy gravel, blast furnace slag and slag should be used to fill the substructure, while the intense weathering soft rocks and collapsible soft rocks shall not be adopted as backfill material.

Table 7.35 Allowable frost heave of substructure in seasonally frozen soil area

Highway class	Allowable frost heave (mm)	
	Cement concrete pavement	Asphalt concrete pavement
Expressway and class-1 highway	≤20	≤40
Class-2 highway	≤30	≤50

Note The allowable frost heave for Class-3 and 4 highway is determined based on the total allowable frost heave of the Class-2 highway with consideration of specific conditions

7.19.5 The embankment height shall follow the specifications in Clauses 3.3.1 and 3.3.2,and the total frost heave shall follow Table 7.35.If not sufficient, the following measures can be taken:

1 Drain the surface water or low the groundwater level.

2 Install frozen prevention cushion, capillary separation layer and drainage layer.

3 Use non-frost heave material or weak frost heave material as backfill material in frozen depth.

4 Use EPS blocks as thermal insulation layer.

Table 7.36 Backfill material for substructures in seasonally frozen soil areas

Substructure type	Frozen zones	Distance of surface water or ground water to the pavement (m)	Frost heave class			
			Upper portion of roadbed	Lower portion of roadbed	Upper portion of embankment	Upper portion of embankment
Fill	Heavy frozen zone	$h_w > 3$	I	I, II, III	–	–
		$h_w \leq 3$	I	I, II	I, II, III	–
	Medium frozen zone	$h_w > 3$	I, II	I, II, III	–	–
		$h_w \leq 3$	I	I, II	–	–
Non-fill or cutting	Heavy frozen zone	$h_w > 3$	I	I	–	–
		$h_w \leq 3$	I	I	–	–
	Medium frozen zone	$h_w > 3$	I	I, II	–	–
		$h_w \leq 3$	I	I	–	–

Note 1. Frost class of soil can be found in Table 7.33

2. In heavy and medium frozen zones, roadbed for Motorways and Class-1 highways use type I soil in which the fine (particle size less than 0.075 mm) content should be not less than 5%
3. In areas with shortage of rocks, the improved backfill material with inorganic binder, slag and stabilization binder may be not limited to this table

7.19.6 Drainage of substructure in seasonally frozen soil areas shall meet the requirements in Chap. 4.The drainage in medium and heavy frozen zones shall also meet the following requirements:

1. Supporting sewer may be installed at wet side soil slopes when groundwater exists at cutting side slopes. oblique drainage holes should be constructed at the groundwater concentrated area.

2 Wide and shallow side ditches should be used for the cutting substructure and the rectangular side ditch with coverage plate is not suitable. When the buried ditch is used, the buried depth shall be not less than 0.25 m below the maximum frozen depth.

3. Drainage sewers shall be used for cutting and fully-frozen embankments. sewers shall be constructed below or outside the side ditch and should not be set in the range of the shoulder.

4 Drainage pipe, water collecting well and sewers shall be buried at least 0.25 m below the local maximum frozen depth. The foundation of the outlet shall be placed below the frost heave line and the outlet of sewers shall be prevented from frozen and kept warm.

Appendix A
Standard Test Method of Dynamic Resilient Modulus of Substructure Soil

A.1 This test method is applicable to determine the dynamic resilient modules of substructure soil and aggregate by dynamic tri-axil apparatus at specified loading conditions.

A.2 Setup of dynamic tri-axil apparatus shall be in accordance with:

1 Tri-axil pressure chamber shall be made of polycarbonate, crylic acid or other suitable transparent material. Lateral pressure fluid should use air as medium.

2 Loading apparatus shall adopt top loading, closed-circuit electrical hydraulic or electric pressure machine that can produce repeated haversine pulse load. The load frequency is 0.1–25 Hz and the applied maximum axial dynamic stress shall not be less than 150 kPa.

3 Computer controlled system shall be used to measure and collect data, which can measure and collect the load and axial deformation of specimen at each load cycle. Pressure gauge, pressure meter or pressure transducer may be used for pressure monitor of tri-axil pressure chamber which shall have a measurement range no less than 200 kPa and measurement accuracy no less than 1.0 kPa. Axial load transducer shall have a measurement range no less than 25 kN and a resolution no less than 5 N.LVDT and other appropriate apparatus may be used displacement transducer which shall have good dynamic response characteristics, a measurement range greater than 6 mm and a resolution no higher than 1% of measurement range.

A.3 Test preparation shall be in accordance with:

1 Specimen preparation shall be in accordance with:

(1) The soil shall be sampled by thin wall.
(2) Particles that are larger than 26.5 mm shall be screened for substructure soil and aggregate with maximum particle size greater than 19 mm. Vibrating or impacting compaction method shall be used to fabricate.
(3) Substructure soil with maximum particle size less than 9.5 mm and the passing ratio of 0.075 mm sieve less than 10% shall be fabricated by vibration compaction.
(4) Substructure soil with maximum particle size less than 9.5 mm and the passing ratio of 0.075 mm sieve no less than 10% shall be fabricated by impacting or static compaction.

W. Wu, *Specifications for Design of Highway Substructure*, JTG Standards and Specifications, https://doi.org/10.1007/978-981-96-2313-6

2 Specimen size shall be in accordance with:

(1) The soil sample length shall not be less than 2 times the diameter of the specimen.
(2) Size of substructure soil and aggregate specimens whose maximum particle size is greater than 19 mm shall be in accordance with diameter by 150 mm ± 2 mm and height by 300 mm ± 2 mm.
(3) Size of substructure soil and aggregate specimens whose maximum particle size is less than 19 mm shall be in accordance with diameter by 100 mm ± 2 mm and height by 200 mm ± 2 mm.

3 Specimen compacted in laboratory shall have a moisture content ± 0.5% of the target moisture content and a compaction degree ± 1.0% of the target compaction degree.

4 Calcium plaster may be used to treat surface defect of hard adhesive specimen (the undrained shear strength is greater than 36 kPa, and the modulus is generally greater than 70 MPa) with a treatment thickness no greater 3 mm.

5 At least three replicates shall be prepared for each test.

A.4 Test procedure shall be in accordance with:

1 A rubber film is sheathed on the specimen to ensure that the seal is airtight.

2 Specimen is placed on the pre-soaked porous permeable stone and the bottom plate. Pre-soaked porous permeable stone and the surface plate are placed on the top of specimen. Pre-soaked filter paper shall be placed between specimen and permeable stone in the case of permeable stone block.

3 The assembled specimen is placed in the center of the base of the tri-axle chamber. The specimen center is aligned with the loading frame center.

4 Install displacement transducer. When the top and bottom top type of measurement devices are used, LVDT or approximate transducer shall be attached to the steel or aluminum rod (between the bottom plate and the surface plate).When the optical deformation measurement device is used, two instructions shall be attached directly to the specimen and each instruction shall be positioned with at least two small pins. When the clamping measurement device is used, the fixture shall be placed at the 1/4 height of the specimen. A soft specimen with undrained shear strength less than 36 kPa shall not be measured by clamping type measurement device.

5 Open the drain valve, connect to the supply pipe and tri-axil chamber, apply pre-confining pressure of 30.0 kPa, and apply the haversine impulse load of 66.0 kPa maximum axial stress for at least 1000 cycles. When the total vertical permanent strain of the specimen reaches 5%, the pre-loading stops. Cause shall be analyzed or specimens shall be reprepared.

6 Adjust the confining pressure and the haversine pulse load to the target values, repeat the load for 100 cycles at 10 Hz frequency. The test collectes the load and deformation curves of the last 5 waveforms, records and calculates the test load, the axial deformation of the specimen, and the dynamic resilient modulus. During loading process, if the total vertical permanent strain of the specimen is more than 5%,the test shall be stopped and the result shall be recorded.

A.5 Calculation of test results shall be in accordance with:

1 Stress Magnitude Shall Be Determined by Eq. (A.5.1):

$$\sigma_0 = \frac{P_i}{A} \tag{A.5.1}$$

where:

σ_0—axial stress amplitude (MPa);

P_i—average amplitude of axial load in the last 5 loading cycles (N);

A—radial cross section area of specimen, taking the mean value of the upper and lower face area of the specimen (mm^2).

2 Strain magnitude shall be determined by Eq. (A.5.2).

$$\varepsilon_0 = \frac{\Delta_i}{l_0} \tag{A.5.2}$$

where:

ε_0—recoverable axial strain amplitude (mm/mm);

Δ_i—the average amplitude of axial deformation that can be recovered in the last 5 loading cycles (mm);

l_0—Measurement distance of displacement transducer (mm).

3 Dynamic resilient modules shall be determined by Eq. (A.5.3):

$$M_R = \frac{\sigma_0}{\varepsilon_0} \tag{A.5.3}$$

where:

M_{R}—the dynamic resilient modulus of substructure soil or aggregate (MPa).

Appendix B
Dynamic Resilient Modulus of Substructure Soil

See (Tables B.1 and B.2).

Table B.1 Reference value of resilient modulus of substructure soil under standard condition

Soil classification	Range of values (MPa)
Gravel (G)	110–135
Gavel containing fined grained soil (GF)	100–130
Silty gravel (GM)	100–125
Clay gravel (GC)	95–120
Sand (S)	95–125
Sand containing fined grained soil (SF)	80–115
Silty sand (SM)	65–95
Clay sand (SC)	60–90
Low liquid limit silt (ML)	50–90
Low liquid limit clay (CL)	50–85
High liquid limit silt (MH)	30–70
High liquid limit clay (CH)	20–50

Note 1. For gravel and sand, the modulus takes high value for high D_{60} (the particle size at the passing ratio of 60%), and the modulus takes low value for low D_{60}

2. For other fine grained soil groups, when the particle content of less than 0.075 mm is large and the plastic index is high, the modulus takes low value; otherwise, the modules takes high value
3. Under the same conditions, the resilient modulus of substructure soil takes smaller value in the light, medium and heavy traffic load and takes higher value at the extra heavy, and extreme heavy traffic load

W. Wu, *Specifications for Design of Highway Substructure*, JTG Standards and Specifications, https://doi.org/10.1007/978-981-96-2313-6

Table B.2 Reference value of resilient modulus of aggregate under standard condition

Aggregate classification	Range of values (MPa)
Graded crushed stone	180–400
Non sieving crushed stone	180–220
Graded gravel	150–300
Natural sand gravel	100–140

Appendix C
Method for Estimating the Equilibrium Moisture of Substructure

C.1 The equilibrium humidity of substructure may be divided three states, wet, medium wet and dry in accordance with substructure humidity, and determine the substructure humidity state in accordance with:

1 The substructure humidity may be classified as wet state when the underground water or longterm surface water has high water level, or the substructure workaround is influenced by the capillary soaking effect of underground water, or the substructure humidity state is controlled by the water level elevation of underground water or long-term surface water.

2 The substructure humidity may be classified as dry state when underground water level is low, or the substructure workaround is above the capillary soaking interface of underground water, or the substructure humidity state is controlled by the climate factors.

3 Humidity of medium wet substructure is influenced by both underground water and climate factors. Substructure workaround is divided as upper and lower parts by the capillary soaking interface of underground water. The lower part is influenced by the capillary soaking effect of underground water and the upper part is influenced by climate factors, as depicted in Fig. C.1.

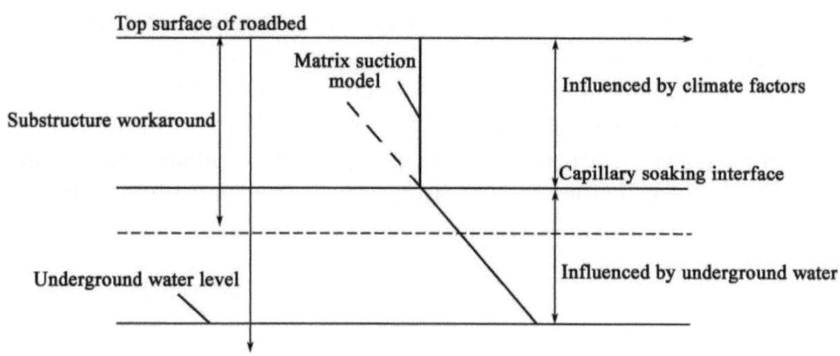

Fig. C.1 Humidity state of medium wet substructure

W. Wu, *Specifications for Design of Highway Substructure*, JTG Standards and Specifications, https://doi.org/10.1007/978-981-96-2313-6

Table C.1 Saturation degree of different substructure soil group at different heights of the groundwater level (%)

Soil classification	The distance of calculation point to groundwater or long-term surface water level (m)						
	0.3	1.0	1.5	2.0	2.5	3.0	4.0
Silty gravel (GM)	69–84	55–69	50–65	49–62	45–59	43–57	–
Clay gravel (GC)	79–96	64–83	60–79	56–75	54–73	52–71	–
Sand (S)	95–80	70–50	–	–	–	–	–
Silty sand (SM)	79–93	64–77	60–72	56–68	54–66	52–64	–
Clay sand (SC)	90–99	77–87	72–83	68–80	66–78	64–76	–
Low liquid limit silt (ML)	94–100	80–90	76–86	83–73	71–81	69–80	–
Low liquid limit clay (CL)	93–100	80–93	76–90	73–88	70–86	68–85	66–83
High liquid limit silt (MH)	100	90–95	86–92	83–90	81–89	80–87	–
High liquid limit clay (CH)	100	93–97	90–93	88–91	86–90	85–89	83–87

Note 1. For sand (SW, SP), the equilibrium humidity takes high value for low D_{60},and modulus takes high value for low D_{60}

2. For other fine grained soil groups, when the particle content of greater than 0.075 mm is large and the plastic index is high, the modulus takes high value; otherwise, the modules takes high value

C.2 Equilibrium humidity of wet substructure may be determined with the saturation degree at different heights of the underground water table specified in Table C.1 in accordance with substructure soil group and underground water level.

C.3 Equilibrium humidity of dry substructure may be determined in accordance with the humidity index TMI of natural zoning and soil group, and shall be in accordance with:

1 The humidity index TMI of natural zoning may refer to Table C.2.

2 The substructure saturation degree can be determined by interpolating the values in Table C.3 in accordance with the TMI value of natural zoning and soil group.

C.4 Equilibrium humidity of medium wet substructure may refer to Fig. C.1.The equilibrium humidity of the upper and lower parts of the substructure workaround are respectively determined. The overall equilibrium humidity is then calculated by the weight of thickness of upper and lower parts of substructure workaround. The equilibrium humidity of the upper substructure workaround above the capillary soaking interface is determined by the soil group and TMI value; the equilibrium humidity of the upper substructure workaround below the capillary soaking interface is determined by the soil group and the distance away from underground water level.

Table C.2 TMI value range of different natural zoning

Natural zoning	Sub-zone		TMI range	Natural zoning	Sub-zone	TMI range
I	I_1		−5.0–8.1	V	V_1	−25.1–6.9
	I_2		0.5–9.7		V_2	0.9–30.1
II	II_1	Heilongjiang	−0.1–8.1		V_{2a}	39.6–43.7
		Liaoning, Jilin	8.7–35.1		V_3	12.0–88.3
	II_{1a}		−3.6–10.8		V_{3a}	−7.6–47.2
	II_2		−7.2–12.1		V_4	−2.6–50.9
	II_{2a}		−1.2–10.6		V_5	39.8–100.6
	II_3		−9.3–26.9		V_{5a}	24.4–39.2
	II_4		−10.7–22.6	VI	VI_1	−15.3–46.3
	II_{4a}		−15.5–17.3		VI_{1a}	−40.5–47.2
	II_{4b}		−7.9–9.9		VI_2	−39.5–59.2
	II_5		−1.7–15.6		VI_3	−41.6
	II_{5a}		−1.0–15.6		VI_4	−19.3–57.2
III	III_1		−21.2–25.7		VI_{4a}	−34.5–37.1
	III_{1a}		−12.6–29.1		VI_{4b}	−2.6–37.2
	III_2		−9.7–17.5	VII	VII_1	−3.1–56.3
	III_{2a}		−19.6		VII_2	−49.4–58.1
	III_3		−19.1–26.1		VII_3	−22.5–82.8
	III_4		−10.8–24.1		VII_4	−5.1–5.7
IV	IV_1		21.8–25.1		VII_5	−20.3–91.4
	IV_{1a}		23.2		VII_{6a}	−10.6–25.8
	IV_2		−6.0–34.8	–	–	–
	IV_3		34.3–40.4	–	–	–
	IV_4		32.0–67.9	–	–	–
	IV_5		45.2–89.3	–	–	–
	IV_6		27.0–64.7	–	–	–
	IV_{6a}		41.2–97.4	–	–	–
	IV_7		16.0–69.3	–	–	–
	IV_{7b}		−5.4–23.0	–	–	–

Table C.3 Saturation degree of different substructure soil at different TMI values (%)

Soil classification	TMI					
	−50	−30	−10	10	30	50
Sand (S)	20–50	25–55	27–60	30–65	32–67	35–70
Silty sand (SM)	45–48	62–68	73–80	80–86	84–89	87–90
Clay sand (SC)						
Low liquid limit silt (ML)	41–46	59–64	75–77	84–86	91–92	92–93
Low liquid limit clay (CL)	39–41	57–64	75–76	86	91	92–94
High liquid limit silt (MH)	41–42	61–62	76–79	85–88	90–92	92–95
High liquid limit clay (CH)	39–51	58–69	85–74	86–92	91–95	94–97

Note 1. Saturation degree of sand is related to D_{60}.It takes low value for high D_{60} (approximating to 2 mm), and takes high value for low D_{60} (approximating to 0.25 mm)

2. The saturation degree of silty sand, clay sand or fine grained soil is related to the content of fine grained soil and plastic index. It takes low value when the content of fine grained soil is high and the plasticity index is high. Otherwise, it takes high value

Appendix D
Range of Humidity Adjustment Coefficient of Substructure Resilient Modulus

D.1 Humidity adjustment coefficient of substructure resilient modulus for wet substructure may refer to Table D.1.

D.2 Humidity adjustment coefficient of substructure resilient modulus for dry substructure may refer to Table D.2.

D.3 The resilient modulus adjustment coefficient of medium wet substructure may be determined by the resilient modulus adjustment coefficients of the upper and lower parts of the substructure workaround. The overall equilibrium humidity of substructure is then calculated by the weight of thickness of upper and lower parts of substructure workaround.

Table D.1 Humidity adjustment coefficient of substructure resilient modulus for wet substructure

Soil classification	Sand	Fine grained soil sand	Silty soil	Clay soil
Surface layer of substructure workaround	0.8–0.9	0.5–0.6	0.5–0.7	0.6–1.0
Bottom layer of substructure workaround	0.5–0.6	0.4–0.5	0.4–0.6	0.5–0.9

Note 1. Adjustment coefficient of resilient modulus of sand takes high value when D_{60} is high and take low value when D_{60} is low

2. Adjustment coefficient of resilient modulus of fine grained soil sand takes low value when the content of fine grained soil is high and the plasticity index is low. Otherwise, it takes high value
3. Adjustment coefficient of resilient modulus of silty soil and clay soil takes low value when the substructure elevation is low. Otherwise, it takes high value

W. Wu, *Specifications for Design of Highway Substructure*, JTG Standards and Specifications, https://doi.org/10.1007/978-981-96-2313-6

Table D.2 Humidity adjustment coefficient of substructure resilient modulus for dry substructure

Soil classification	TMI					
	−50	−30	−10	10	30	50
Sand (S)	1.30–1.84	1.14–1.80	1.02–1.77	0.93–1.73	0.86–1.69	0.8–1.64
Silty sand (SM)	1.59–1.65	1.10–1.26	0.83–0.97	0.73–0.83	0.70–0.76	0.70–0.76
Clay sand (SC)						
Low liquid limit silt (ML)	1.35–1.55	1.01–1.23	0.76–0.96	0.58–0.77	0.51–0.65	0.42–0.62
Low liquid limit clay (CL)	1.22–1.71	0.73–1.52	0.57–1.24	0.51–1.02	0.49–0.88	0.48–0.81

Note 1. Resilient modulus adjustment coefficient of sand is related to D_{60}.It takes low value for high D_{60} (approximating to 2 mm), and takes high value for low D_{60} (approximating to 0.25 mm)

2. The saturation degree of silty sand, clay sand or fine grained soil is related to the content of fine grained soil and plastic index. It takes low value when the content of fine grained soil is high and the plasticity index is high. Otherwise, it takes high value

Appendix E
Rock Classification for Rock Side Slope

E.1 The intactness of rock mass shall be determined by the development of structural plane, structural type and intactness coefficient specified in Table E.1 .The intactness coefficient shall be calculated by Eq. (E.1).

$$Kv = \left(\frac{v_R}{v_P}\right)^2 \tag{E.1}$$

where:

K_v—intactness coefficient of rock mass;

v_R—propagation velocity of elastic longitudinal wave in rock mass (km/s);

v_P—propagation velocity of elastic longitudinal wave in rock block (km/s).

Table E.1 Classification of intactness of rock mass

Intactness of rock mass	Development of structural plane	Structural type	Intactness coefficient K_v
Integrated	Having one or two groups of structural planes based on structural joints, layered structure and sealed type	Global structure with giant block	>0.75
Comparatively integrated	Having two or three groups of structural planes based on structural joints and a layered structure; most cracks are sealed, some cracks are slightly open and a few cracks are filled	Block structure, layered structure, and fragmented mosaic structure	0.35–0.75
Incomplete	Having more than 3 groups of structural planes, near the fault influenced by the tectonic action; most cracks are open, filled with filling and thick	Fragmented structure and granular structure	<0.35

Note Fragmented mosaic structure is a structure that has larger pieces of fractured fragments that are tightly interlocked and have good stability

W. Wu, *Specifications for Design of Highway Substructure*, JTG Standards and Specifications, https://doi.org/10.1007/978-981-96-2313-6

E.2 Rock classification shall be carried out in accordance with intactness of rock mass, binding of structural planes, occurrence of structural planes and self-stability of vertical slope, and shall be in accordance with Table E.2.

Table E.2 Rock classification for rock side slope

Rock type of side slope	Determine conditions			
	Intactness of rock mass	Binding of structural planes	Occurrence of structural planes	Self stability of vertical slope
I	Integrated	Good or common binding of structural planes	The composite dip angle of outwardly inclined structural planes or different outwardly inclined structural planes is greater than 75°or less than 35°	30 m high slope has long-term stability with occasional rockfalls
II	Integrated	Good or common binding of structural planes	The composite dip angle of outwardly inclined structural planes or different outwardly inclined structural planes is in the range of 35°–75°	15 m high slope is stable;15–30 m high slope lacks of stability
	Integrated	Poor binding of structural planes	The composite dip angle of outwardly inclined structural planes or different outwardly inclined structural planes is greater than 75°or less than 35°	
	Relatively integrated	Good, common or poor binding of structural planes	The composite dip angle of outwardly inclined structural planes or different outwardly inclined structural planes is less than 35°,having inward structure plane	Slope with partial collapsing
III	Integrated	Poor binding of structural planes	The composite dip angle of outwardly inclined structural planes or different outwardly inclined structural planes is in the range of 35°–75°	8 m high slope is stable,15 m high slope lacks of stability
	Relatively Integrated	Good or common binding of structural planes	The composite dip angle of outwardly inclined structural planes or different outwardly inclined structural planes is in the range of 35°–75°	8 m high slope is stable,15 m high slope lacks of stability

(continued)

Table E.2 (continued)

Rock type of side slope	Determine conditions			
	Intactness of rock mass	Binding of structural planes	Occurrence of structural planes	Self stability of vertical slope
	Relatively integrated	Poor binding of structural planes	The composite dip angle of outwardly inclined structural planes or different outwardly inclined structural planes is greater than 75°or less than 35°	
	Relatively integrated (fragmentation inserted)	Good or common binding of structural planes	Structural planes without obvious rules	
IV	Relatively integrated	Poor or very poor binding of structural planes	Outwardly inclined structural planes with layers and dip angles that are mostly in the range of 35°–75°	8 m high slope is unstable
	Incomplete (disintegrated, fragmented)	Very poor binding between broken rocks	–	

Note 1. Slope rock classification excludes slopes that are controlled by a weak structure plane and dumping collapse type slopes

2. When Class-1 rock is a soft rock or relatively soft rock, it shall be reduced to Class-2
3. When underground water is well developed, Class-2 and 3 rocks may be reduced one level depending on the particular case
4. Highly weathered rock and extremely soft rock may be classified as Class-4 rock
5. Outward inclined structural plane used in the table is the structural plane with angle between dip direction and slope direction less than 30°

Appendix F
Monitoring Content and Items of Substructure

See (Tables F.1, F.2, and F.3).

Table F.1 Cutting slope and landslide monitoring

Monitoring content		Monitoring method	Monitoring purposes
Earth surface monitoring	Horizontal displacement Monitoring	Total station, EDM	Observing displacements and deformation development of earth surface
	Vertical deformation monitoring	Level	
	Crack monitoring	Stake, ruler or crackmeter	Observing the development of cracks
Underground displacement monitoring		Inclinometer	Probing displacement of the underground rock mass relative to stable rock; confirming and ascertaining the structural characteristics of rock that is displacing; determining the depth of the potential sliding surface; judging the main slide direction; performing a quantitative analysis evaluation of slope stability conditions; evaluating engineering effects of slope reinforcement
Underground water monitoring		Manual measurement	Observing the relationship between the variation of ground water levels and rainfall; evaluating the effectiveness of slope drainage measures

(continued)

W. Wu, *Specifications for Design of Highway Substructure*, JTG Standards and Specifications, https://doi.org/10.1007/978-981-96-2313-6

Table F.1 (continued)

Monitoring content	Monitoring method	Monitoring purposes
Deformation and stress of supporting and retaining structures	Inclinometer, layer settlement gauge pressure cells, and reinforcing steel stress meter	Deformation observations of supporting and retaining structure; pressure observations between supporting and retaining structure and rock and soil mass

Table F.2 High embankment stability and settlement observation

Monitoring items	Instruments	Monitoring purposes
Earth surface horizontal displacement and upheaval	Earth surface horizontal displacement stake (side stake)	Monitoring stability to ensure construction safety and embankment stability
Underground soil stratified horizontal displacement	Underground horizontal displacement gauge (inclinometer tube)	Monitoring stability and researching; measuring layer displacement; determining the location of shear failure when necessary
Settlement at top of embankment	Ground settlement meter (settlement plate or settlement stake)	Monitoring settlement after construction; forecasting settlement trends; determining pavement construction time

Table F.3 In situ monitoring content and items in prestressed anchorage projects

Stages of anchor		Monitoring purposes	Monitoring items	Monitoring method and requirement
Construction stage	Prestressed anchor	Construction safety Construction quality	Prestress	Project shall be made, including anchor tension force and prestress force loss. Anti-pull method should be used. Embedded instrument method may be used
			Anchor displacement	Project shall be made. Conventional method of displacement monitoring should be used

(continued)

Table F.3 (continued)

Stages of anchor		Monitoring purposes	Monitoring items	Monitoring method and requirement
			Deep displacement of rock and soil body	Project shall be made. Inclinometer method may be used
			Anchor length	Project may be made. Nondestructive testing method should be used
			Grouting fullness	Project may be made. Nondestructive testing method should be used
Project operational stage	Prestressed anchor	Working status	Prestress	Project shall be made. Anti-pull method should be used. Embedded instrument method may be used
			Anchor displacement	Project should be made. Conventional method of displacement monitoring should be used
			Deep displacement of rock and soil body	Project may be made. Embedded instrument method may be used

Appendix G
Material Strength Requirements of Roadbed Drainage, Protection and Retaining Structures

See (Tables G.1 and G.2).

Table G.1 Material strength requirements of drainage structures

Material type	Minimum strength requirements		Range of application
	Non-frozen zone, light frozen zone	Intermediate frozen zone, heavy frozen zone	
Rubble	MU30	MU30	Ditch bottom, ditch sides paving
Cement mortar	M7.5	M10	Masonry, plaster, joint
Cement concrete	C20	C25	Concrete member
	C15	C15	Concrete foundation

Note Light frozen zone is the area where frozen index is lower than 800.Intermediate frozen zone is the area where frozen index is between 800 and 2000.Heavy frozen zone is the area where frozen index is higher than 2000

W. Wu, *Specifications for Design of Highway Substructure*, JTG Standards and Specifications, https://doi.org/10.1007/978-981-96-2313-6

Table G.2 Material strength requirements of protection and retaining structures

Material type	Minimum strength grade		Range of application
	Non-frozen zone, light frozen zone	Intermediate frozen zone, heavy frozen zone	
Rubble	MU30	MU40	Slope protection, facing wall, retaining wall
Cement mortar	M7.5	M10	Slope protection, facing wall, retaining wall
	M10		Shotcrete protection
Cement concrete	C15	C20	Shotcrete, retaining wall foundation, locking wellhead of anti-slide pile, wall protection
	C20	C25	Slope protection, various retaining walls, soil nailing panel
	C30	C30	Anti-slide pile, anchor support, Lattice (frame) beam, ground beam, single anchor pier

Appendix H
Design and Calculation of Retaining Wall

H.1 Loads shall meet the following specifications:

1 The design and calculation of retaining wall applied in this specification shall be based on the partial factor method of limit state design.

2 The general expression for the ultimate limit state design of the retaining wall element may be as follows:

$$\gamma_0 S \leq R \tag{H.1.1}$$

$$R = R\left(\frac{R_k}{\gamma_f}, \alpha_d\right) \tag{H.1.2}$$

where:

γ_0—structural important coefficient, determined according to those specified in Table H.1;

S—design value of action (or load) combination effect;

$R(\cdot)$—structural resistance strength of the retaining wall;

R_k—strength standard value of the resistance material;

γ_f—partial coefficient of the structural material or geotechnical properties;

α_d—design geometric parameters of the structure or element, which may be determined as equal to standard geometric parameters when reliable data is not available.

3 The actions (or loads) exerted on the retaining wall may be listed in Table H.2 according to their characteristics.

Table H.1 Structural important coefficient γ_0

Height of retaining wall	Class of highway	
	Motorway, Class-1 highway	Class-2 or lower class highway
≤5.0 m	1.0	0.95
>5.0 m	1.05	1.0

W. Wu, *Specifications for Design of Highway Substructure*, JTG Standards and Specifications, https://doi.org/10.1007/978-981-96-2313-6

Table H.2 Load classification

Classification of actions (or loads)		Name of actions (or loads)
Permanent effects (or loads)		Gravity of retaining wall
		Gravity of fill (including the fill above the foundation)
		Lateral pressure of fill
		Permanent load on the top of retaining wall
		Effective load between the top of wall and second fracture plane
		Buoyancy from the calculated level of water and hydrostatic pressure
		Pre-applied load
		Shrinkage and creep of concrete
		Effect induced from the displacement of foundation
Variable actions (or loads)	Basic variable actions (or loads)	Soil lateral pressure induced from vehicle load
		Pedestrian load and soil lateral pressure induced from pedestrian load
	Other variable actions (or loads)	Hydrodynamic pressure caused by the falling of water level
		Pressure of flowing water
		Wave pressure
		Frost heaving pressure and ice pressure
		Temperature effect
	Construction load	Temporary load related to various retaining wall construction
Accidental actions (or loads)		Earthquake action
		Landslide and debris flow action
		Collision force of the vehicle on guard rail at the top of retaining wall

4 Loading combinations shall meet the following specifications:

(1) Only the permanent action (or loads) and basic variable actions (or loads) may be calculated in the design for the retaining wall located in the general area.
(2) Other variable actions (or loads) and accidental actions (or loads) shall also be calculated if the retaining wall is located in a soaking area, frost heaving area

Table H.3 Common load combinations

Combination	Name of actions (or loads)
I	Combination of gravity of retaining wall, permanent load on the top of retaining wall, gravity of fill, lateral pressure of fill and other permanent loads
II	Combination I combines with basic variable loads
III	Combination II combines with other variable loads and accidental loads

Note 1. Flood effect and earthquake effect could not be considered together
2. Frost heaving pressure and ice pressure could not be considered together with the pressure of flowing water or waves
3. Vehicle loads and earthquake effects could not be considered together

or areas where the seismic peak ground acceleration is equal to or larger than 0.2 g.

(3) The load combinations may be carried out according to Table H.3.

5 For retaining walls subject to earthquake force, the requirements of the *Code for Seismic Design of Highway Engineering* (JTG B04) shall be met.

6 For anti-slide retaining walls with an obvious sliding surface, the load calculations shall meet the requirements stated in Sects. 5.7 and 7.2. In the design of the retaining wall in a debris flow section, the calculations shall meet the requirements stated in Sect. 7.5.

7 The hydrostatic pressure on both sides of the soaking retaining wall and the hydrodynamic pressure on the back of this kind of wall may be neglected when rock blocks or coarse grain soil is filled behind.

8 The buoyancy shall be defined according to the principles below:

(1) The calculated level of water shall be fulfilled to 100% when the subsoil consists of gravelly soil, debris soil or joint-developed rock.

(2) The calculated level of water shall be reduced to 50% when the subsoil consists of rock.

9 The active earth pressure on the back of the retaining wall may be calculated through the Coulomb Theory. A test of soil texture for the fill behind the retaining wall shall be conducted in order to determine the physic indexes of the fill. The angle of internal friction φ may be selected from Table H.4 when reliable test data is not available.

10 The passive earth pressure acting on the front face of the retaining wall may not be counted unless the foundation is deeply embedded, not scoured by flowing water, not affected by other disturbances and located in an area with stable ground. Partial factors of effects shall be counted according to what is specified in Table H.5.

11 Soil lateral pressure on the back of the retaining wall induced from vehicle load may be defined as equal to the pressure induced by a layer of soil with a constant thickness. The thickness may be calculated according to the following equation:

$$h_0 = \frac{q}{\gamma} \tag{H.1.3}$$

Table H.4 Angle or equivalent angle of internal friction of fill (°)

Type of fill		Equivalent angle of internal friction φ_0(°)	Angle of internal friction φ (°)	Unit weight (kN/m^3)
Cohesive soil	Height of retaining wall $H \leq 6$ m	35–40	–	17–18
	Height of retaining wall $H > 6$ m	30–35	–	
Broken stone, block stone that are not prone to weathering		–	45–50	18–19
Big cobble stone, debris soil, broken rock that are not prone to weathering		–	40–45	18–19
Small cobble stone, gravel, coarse sand, chip		–	35–40	18–19
Medium sand, fine sand, sandy soil		–	30–35	17–18

Note The unit weight of fill may be amended moderately according to the test data. The unit weight of fill that is below the calculated level of water is defined as its buoyancy unit weight

where:

h_0—thickness of the converted soil layer (m);

q—surcharge load converted equally from the vehicle load; the degree is valued at 20 kN/m^2 when the height of retaining wall is less than 2 m; when the height is larger than 10 m, it is valued at 10 kN/m^2;its value is obtained through linear interpolation when the height is between 2 and 10 m; the strength of pedestrian load on wall top or backfill of backside of wall is 3 kN/m^2;the strength of horizontal thrusting force on top of hand-rail of retaining wall is 0.75 kN/m^2;the strength of vertical force on hand-rail is 1 kN/m^2;

γ—unit weight of fill (kN/m^3).

12 When the retaining wall may be designed according to the ultimate limit state, the partial factors of common actions (or loads) are listed in Table H.5.

H.2 Foundation design and stability calculation shall meet the following specifications:

1 The eccentricity of the resultant force on the bottom of the foundation may be calculated according to the following formula:

$$e_0 = \frac{M_{\rm d}}{N_{\rm d}} \tag{H.2.1}$$

where:

$N_{\rm d}$—combination design value of vertical load on the bottom of foundation (kN/m);

$M_{\rm d}$—combination design value of the moment about the centroid of the bottom of the foundation (MPa).

Table H.5 Partial factors of actions (or loads) in ultimate limit state

Case	The increment of load is favorable to the retaining wall		The increment of load is unfavorable to the retaining wall	
Combination	I, II	III	I, II	III
Vertical permanent load γ_G	0.90		1.20	
Active earth pressure induced from permanent load, vehicle load or pedestrian load γ_{Q1}	1.00	0.95	1.40	1.30
Passive earth pressure γ_{Q2}	0.30		0.50	
Buoyancy γ_{Q3}	0.95		1.10	
Hydrostatic pressure γ_{Q4}	0.95		1.05	
Hydrodynamic pressure γ_{Q5}	0.95		1.20	

2 When calculating the foundation of retaining wall, the partial factors in the determinations of structural resistance strength design values of all kinds of actions (or loads) is prescribed as 1. Excluding passive earth pressure, the partial factor of passive earth pressure γ_{Q2} is prescribed to be 0.3.

3 The compressive stress on the bottom of the foundation shall be calculated according to Eq. (H.2.2).When the retaining wall is located on rock, the compressive stress may be calculated according to Eqs. (H.2.3) and (H.2.4).The eccentricity e_0 of the resultant force on the bottom of the foundation shall not be greater than *B*/6 for soil foundation and *B*/4 for rock foundation. The compressive stress on the bottom of the foundation shall not be greater than the allowable stress $[\sigma_0]$ at the bottom of the foundation. The allowable stress at the bottom of the foundation may be adopted according to values specified in the *Code for Design of Ground Base and Foundation of Highway Bridges and Culverts* (JTG D63).When checking that the loading combination III or $[\sigma_0]$ is greater than 150 kPa, the allowable stress at the bottom of the foundation may be increased by 25%.

$$|e_0| \le \frac{B}{6'}\sigma_{1,2} = \frac{N_d}{A}\left(1 \pm \frac{6e_0}{B}\right) \tag{H.2.2}$$

$$e_0 > \frac{B}{6'}\sigma = \frac{2N_d}{3\alpha_1}, \sigma_2 = 0 \tag{H.2.3}$$

$$\alpha 1 = \frac{B}{2} - e_0 \tag{H.2.4}$$

where:

σ_1—compressive stress at the toe of the retaining wall (kPa);
σ_2—compressive stress at the heel of the retaining wall (kPa);

Table H.6 Friction coefficient between the bottom of foundation and the subsoil μ

Category of subsoil	Friction coefficient μ	Category of subsoil	Friction coefficient μ
Soft plastic clay	0.25	Debris soil	0.50
Hard plastic clay	0.30	Soft rock	0.40–0.60
Gravelly soil, sandy clay, semi-dry-hard clay	0.30–0.40	hard rock	0.60–0.70
Sandy soil	0.40		

B—width of the bottom of the foundation (m), is also the length of the hypotenuse if the bottom of the foundation is inclined;

A—area per unit length, equal to $B \times 1$ (m^2) when the bottom of the foundation is rectangular foundation.

4 The embedded depth of the foundation shall meet Eq. (H.2.5).the anti-sliding stability coefficient shall be calculated according to Eq. (H.2.6).

$$[1.1G + \gamma_{Q1}(E_y + E_x\tan\alpha_0) - \gamma_{Q2}E_p\tan\alpha_0]$$
$$\mu + (1.1G + \gamma_{Q1}E_y)\tan\alpha_0 - \gamma_{Q1}E_x + \gamma_{Q2}E_p > 0 \tag{H.2.5}$$

$$K_c = \frac{[N + (E_x - E'_p)\tan\alpha_0]\mu + E'_p}{E_x - N\tan\alpha_0} \tag{H.2.6}$$

where:

G—total gravity (kN) acting above the bottom of the foundation. The soaked part of the retaining wall shall include buoyancy;

E_y—vertical component of active earth pressure on the back of the retaining wall (kN);

E_x—horizontal component of active earth pressure on the back of the retaining wall (kN);

E_p—horizontal component of passive earth pressure on the front of the retaining wall (kN); if retaining wall is soaked, $E_p = 0$;

E'_p—0.3 times of horizontal component of passive earth pressure on the front of the retaining wall (kN);

α_0—tilt angle of the bottom of the foundation (°); $\alpha_0 = 0$ if the bottom is horizontal;

γ_{Q1},γ_{Q2}—partial factors of active/passive earth pressure on the back/front of the retaining wall; these factors may be selected from Table H.5;

μ—is the friction coefficient between the bottom of the foundation and the subsoil; this coefficient may be selected from Table H.6 when reliable test data is not available.

5 The overturning stability equation of the retaining wall shall meet the requirements of Eq. (H.2.7).The anti-overturning stability coefficient shall be calculated according to Eq. (H.2.8).

$$0.8GZ_G + \gamma_{Q1}(E_yZ_x - E_yZ_y) + \gamma_{Q2}E_pZ_p > 0 \tag{H.2.7}$$

Table H.7 Anti-sliding and anti-overturning stability coefficients

Load case	Checking item	Stability coefficient	
Load combination I, II	Anti-sliding	K_c	1.3
	Anti-overturning	K_0	1.5
Load combination III	Anti-sliding	K_c	1.3
	Anti-overturning	K_0	1.3
Construction stage checking	Anti-sliding	K_c	1.2
	Anti-overturning	K_0	1.2

$$K_0 = \frac{GZ_G + E_y Z_x + E'_p Z_p}{E_x Z_y} \tag{H.2.8}$$

where:

Z_G—distance between the resultant force of the gravity of the wall body, the foundation, the fill above the foundation, the vertical component of other load forces on top of the wall and the toe of the wall (m);

Z_x—distance between the vertical component of active earth pressure on the back of the retaining wall and the toe of the wall (m);

Z_y—distance between the horizontal component of active earth pressure on the back of the retaining wall and the toe of the wall (m);

Z_p—distance between the horizontal component of passive earth pressure on the front of the retaining wall and the toe of the wall (m).

6 In the checking of anti-sliding stability and anti-overturning stability for the retaining wall with heights in the realm specified in the *Specifications*, the stability coefficients shall not be less than the values specified in Table H.7.

7 The global stability of the retaining wall subsoil and fill shall be checked if the retaining wall is on poor subsoil, with inclined bedrock under the surface soil or set on a slope. The stability coefficient shall not be less than 1.25.

H.3 Design and calculation of gravity and semi-gravity retaining wall shall meet the following specifications:

1 The effect (or loads) calculations of gravity or semi-gravity retaining walls shall be in accordance with Clause H.1 of the *Specifications.*

2 The foundation design and stability calculation for gravity or semi-gravity retaining walls shall meet the requirements specified in Clause H.2 of the *Specifications.*

3 The adoption of material strength for the body of gravity or semi-gravity retaining walls may be in compliance with the requirements of the *Code for Design of Highway Masonry Bridges and Culverts* (JTG D61). If necessary, the shear stress of the walls shall be checked.

4 When a gravity retaining wall is designed according to the ultimate limit state, the modular design value of action (or load) effect in some certain loading combination may be calculated according to Eq. (H.3.1).The partial resistance factor of the

Table H.8 Partial resistance factors of masonry element or material γ_f

Type of masonry	Force condition	
	In compression	In bending, shear, or tension
Stone	1.85	2.31
Rubble masonry, rubble concrete masonry	2.31	2.31
Block stone, coarse dressed stone, concrete precast block, brick masonry	1.92	2.31
Concrete	1.54	2.31

Table H.9 Load effect combination factor

Loading combination	Ψ_{ZL}	Loading combination	Ψ_{ZL}	Loading combination	Ψ_{ZL}
I, II	1.0	Construction load	0.7	III	0.8

masonry element or material γ_f is adopted from Table H.8.

$$S = \psi_{ZL}(\gamma_G \sum S_{Gik} + \sum \gamma_{Qi} S_{Qik}) \tag{H.3.1}$$

where:

S—action (or load) effect combination design value;

γ_G, γ_{Qi}—partial effect (or load) factors; they are adopted from Table H.5 of this specification;

S_{Gik}—standard value of the effect induced by the vertical permanent load i;

S_{Qik}—standard values of the effects induced by soil lateral pressure, buoyancy, hydrostatic pressure, and other actions (or loads);

ψ_{ZL}—load effect combination factor; it shall be adopted from Table H.9.

5 When a retaining wall member is subjected to axial or eccentrical compression, its normal section strength and stability is checked according to Eqs. (H.3.2) and (H.3.3).As far as the eccentric compression member is concerned, the longitudinal stability of its in-plane bending shall be checked while the stability of its out-of-pane bending shall also be checked as an axial compression member.

For strength checking:

$$\gamma_0 Nd \le \frac{\alpha_k A R_\alpha}{\gamma_f} \tag{H.3.2}$$

For stability checking:

$$\gamma_0 Nd \le \frac{\psi_k \alpha_k A R_\alpha}{\gamma_f} \tag{H.3.3}$$

where:

N_d—combination design value of axial forces on the normal section (kN);

γ_0—structural important coefficient that can be adopted according to Clause H.1 of the *Specifications*;

γ_f—partial resistance factor of masonry member or material that can be adopted from Table H.8;

R_α—nominal ultimate compressive strength of material (kN);

A—calculated section area of the element of retaining wall (m^2);

α_k—eccentric influence factor of axial force that can be calculated according to Eq. (H.3.4);

ψ_k—in-plane longitudinal bending factor for eccentric compression member that can be calculated according to Eq. (H.3.6); an axial compression member may be adopted from Table H.11.

$$\alpha_k = \frac{1-256\left(\frac{e_0}{B}\right)^8}{1+12\left(\frac{e_0}{B}\right)^2} \tag{H.3.4}$$

where:

e_0—axial force eccentricity that can be calculated by equation (H.3.5); where the body or foundation of the retaining wall is of masonry section, e_0 shall comply with the provisions of Table H.12;

B—width of the calculated section of the retaining wall element (m).

$$e0 = \left|\frac{M_0}{N_0}\right| \tag{H.3.5}$$

where:

M_0—total moment caused by a certain action (or load) combination on the calculated section (the moment here rotates around the centroid of the section) ($kN \cdot m$);

$N0$—total force caused by a certain action (or load) combination on the calculated section (kN).

$$\psi_k = \frac{1}{1+\alpha_s\beta_s(\beta_s-3)\left[1+16\left(\frac{e_0}{B}\right)^2\right]} \tag{H.3.6}$$

$$\beta_s = \frac{2H}{B} \tag{H.3.7}$$

where:

H—height of the retaining wall (m);

α_s—coefficient related to material (see Table H.10).

The other symbols have unchanged meanings (Table H.11).

6 The eccentricity of axial force in the gravity retaining wall, e_0, shall be in accordance with those specified in Table H.12.

Table H.10 Values of a_s

Type of masonry	Strength grade of mortar in grouted masonry			Concrete
	M10, M7.5, M5	M2.5	M1	
Values of a_s	0.002	0.0025	0.004	0.002

Table H.11 Longitudinal bending factors for axial compression member Ψ_k

$2H/B$	Concrete member	Strength grade of mortar in grouted masonry	
		M10, M7.5, M5	M2.5
≤3	1.00	1.00	1.00
4	0.99	0.99	0.99
6	0.96	0.96	0.96
8	0.93	0.93	0.91
10	0.88	0.88	0.85
12	0.82	0.82	0.79
14	0.76	0.76	0.72
16	0.71	0.71	0.66
18	0.65	0.65	0.60
20	0.60	0.60	0.54
22	0.54	0.54	0.49
24	0.50	0.50	0.44
26	0.46	0.46	0.40
28	0.42	0.42	0.36
30	0.38	0.38	0.33

Table H.12 Allowable eccentricity for the axial resultant force in masonry structure e_0

Load combination	Allowable eccentricity	Load combination	Allowable eccentricity
I, II	$0.25B$	Construction load	$0.33B$
III	$0.3B$		

Note B is the length of one side of the rectangular calculated section which is parallel to the in-plane moment on the section

7 The tensile side of a concrete section shall be provided with reinforcement, with no less than 0.05% reinforcement ratio. The allowable eccentricity in Table H.12 may be increased by $0.05B$. Furthermore, when the reinforcement ratio is larger than what is specified in Table H.13, the member need to be designed as a reinforced concrete member while the restriction on eccentricity does not apply.

H.4 The design calculation of the ultimate limit state, checking calculation of the serviceability limit state, and structural requirements of cantilever and counterfort

Table H.13 The minimum tensile reinforcement ratio of the section designed as reinforced concrete member

Trademark of steel bar (type)	Minimum reinforcement ratio (%)	
	Single-side-reinforced section	All-reinforced section
Q235 (Grade I)	0.20	0.50
HRB335, HRB400 (Grade II and III)	0.20	0.50

Note The calculation of the minimum reinforcement ratio considers the total cross section area

retaining walls shall comply not only with the *Specifications* but also with the requirements of the *Code for Design of Highway Reinforced Concrete and Prestressed Concrete Bridges and Culverts* (JTG D62).

1 For the design of cantilever and counterfort retaining walls, the requirements for the foundation design and stability calculation specified in Clause H.2 of the *Specifications* shall be met.

2 The calculation for actions (or loads) on the retaining wall shall be in compliance with the requirements in Clause H.1 of the *Specifications.* The friction between the fill and surfaces of the stem and heel may be neglected when calculating the earth pressure on the stem and heel.

3 The earth pressure on the front of the retaining wall may be neglected when calculating the global stability of the retaining wall and stem; the weight above the bottom plate shall be taken into account when calculating the toe plate internal forces.

4 Each part of a cantilever retaining wall shall be analyzed as a cantilever beam. The adoption of partial factors of actions (or loads) shall be in compliance with the requirements in Clause H.1 of the *Specifications.* When the normal stress on the bottom of the foundation is taken as a vertical load, the partial factor may be defined as the same as that for a permanent vertical load.

5 The toe plate of the counterfort retaining wall may be analyzed as a cantilever beam while the heel plate may be analyzed as a continuous plate supported on counterforts without the effect of constraint by the vertical wall plate. The counterfort may be analyzed as a cantilever T beam. The bending moment along the route direction in the vertical wall plate may be analyzed as a continuous beam supported on counterforts.

6 The distribution of actions (or loads) on the vertical wall plate along the height of the wall may be assumed to have a ladder-shaped profile [as shown in Fig. H.1a]. Figure H.1b shows the vertical moment distribution along the height of the wall; the vertical moment along the direction of the road line takes a step-shaped distribution [as shown in Fig. H.1c]. The bending moment along the direction of the road line in the vertical wall plate may be calculated as a continuous beam supported on counterforts.

M_{mid}-the moment in the middle of the plate span; -the height of the vertical wall plate; σ_{H}-the normal earth pressure caused by fill behind the foot of the vertical wall plate; l-the clear distance between adjacent counterforts.

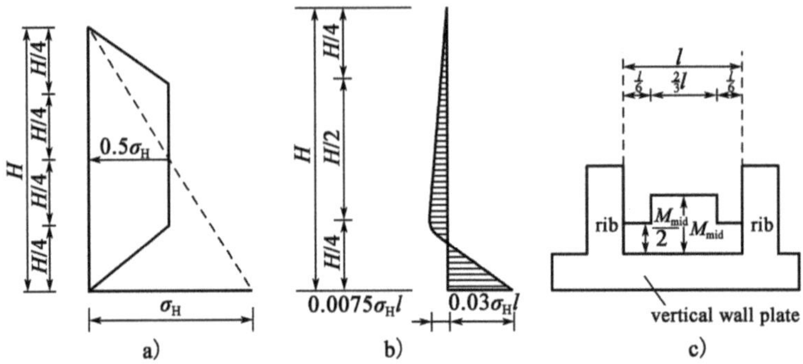

Fig. H.1 Distributions of load and moment

H.5 The design calculation of ultimate limit state, checking calculation of serviceability limit state, and structural requirements for the design of reinforced concrete member of the anchored retaining wall by tie rods shall comply not only with the *Specifications* but also with the requirements of the *Code for Design of Highway Reinforced Concrete and Prestressed Concrete Bridges and Culverts* (JTG D62).

1 The calculation for actions (or loads) on the anchored retaining wall by tie rods shall be in compliance with the requirements in Clause H.1 of the *Specifications*.

2 When the wall is divided into multi-tiers, the active earth pressure on the back of each wall tier may be calculated using the extended wall back method.

3 The design calculation for rib column shall be in accordance with the following requirements:

(1) The actions (or loads) on the rib column shall be taken as the actions (or loads) on the wall plate between the two center lines of adjacent plate spans.
(2) Depending on the geological conditions, bearing capacity and embedded length of the rib column foundation, the lower end of the rib column may be considered as a hinge or a free end without any constraints by the foundation soil. The internal forces and support reactions of the rib column shall be calculated as a simply-supported beam or a continuous beam supported by the anchor tie rods.
(3) In the calculation of section strength and reinforcement of rib column, the combined design values of internal forces shall be used. The action (load) partial factors specified in Sect. H.1 shall be used.
(4) For precast rib columns, the section strength shall be checked for the loads generated in transportation, lifting, and construction as well as the uneven load caused by the disproportion of force among the anchoring tie rods.
(5) Fabricated retaining plate may be analyzed as a simply-supported beam supported by rib columns. The calculation span length is equal to the clear distance between adjacent rib columns plus the lap lengths at both ends of the plate.

4 The calculations for the internal forces of a cast-in-place anchored retaining wall by tie rods may be analyzed as a continuous beam with unit width in both vertical and

Fig. H.2 Distributions of permanent earth pressure

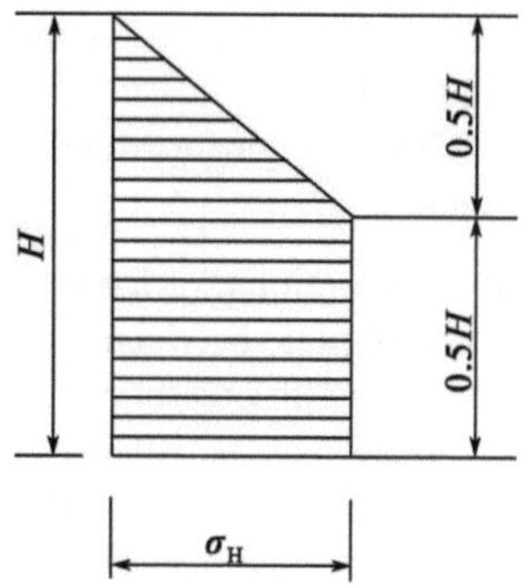

horizontal directions. The earth pressure on the back of the wall plate is considered the load on the vertical unit width continuous beam while the load on the horizontal unit width continuous beam is the maximum earth pressure on the part of the plate where the horizontal unit width beam is located.

H.6 The design calculation of ultimate limit state, calculation check of serviceability limit state, and structural requirements for the reinforced concrete members of the anchored bulkhead retaining wall shall comply not only with the *Specifications* but also with the requirements of the *Code for Design of Highway Reinforced Concrete and Prestressed Concrete Bridges and Culverts* (JTG D62).

1 In the combination with action (or load) effects of the calculations for designing the reinforced concrete members of the anchored bulkhead retaining wall, the values of structural important coefficient γ_0 shall be determined according to Sect. H.1 of this specification.

2 The earth pressure due to permanent loads distributes on the anchored bulkhead or the retaining plate of anchored bulkhead retaining wall as shown in Fig. H.2. The horizontal earth pressure shall be calculated according to Eq. (H.6.1).

$$\sigma_H = \frac{1.33E_x}{H}\beta \tag{H.6.1}$$

where:

σ_{H}—horizontal earth pressure caused by permanent loads (kPa);

E_x—horizontal component of the active earth pressure calculated through the Coulomb theory on the back of unit length wall (kN/m);

H—height of the wall or the sum of the heights of upper and lower wall if the wall is divided into two levels (m);

β—amplification coefficient of the earth pressure, ranging from 1.2 to 1.4;the earth pressure caused by the vehicle load is not amplified.

3 The fold-line sliding surface method or whole wall method may be used to check the global sliding stability of the anchored retaining wall. The sliding stability coefficient shall not be less than 1.8.When determining sliding stability, the adopted calculation value shall be the most unfavorable combination of whether or not there

is an additional load on top of the wall and whether or not the earth pressure is amplified.

4 The design for rib column shall be in accordance with the following requirements:

(1) The actions (or loads) on the rib column shall be the actions (or loads) on the wall plate between the two centerlines of the adjacent plate spans.
(2) The earth pressure acting on the rib column transferred from the retaining wall plate may be calculated using a simply-supported beam or continuous beam according to the number of rows of tie rods and the way in which the rib column and its foundation are connected.

5 The design for tie rod shall be in accordance with the following requirements:

(1) The tie rods in the top row shall be at least 1 m beneath the fill surface. The tie rod may be set at a downward inclination if the embedded length of the anchored bulkhead is not enough. The slope of the inclined tie rod should be at 10°–15°.
(2) The length of the tie rod shall be determined according to the requirements for the global stability of the retaining wall and the length of the tie rod in the bottom row extending beyond the fracture surface caused by active pressure load. In addition, the tie rod length shall not be less than 3.5 times the height of the anchored bulkhead. The length of the tie rods in the top row shall not be less than 5 m.
(3) The diameter of a single steel tie rod without considering rust residue is calculated according to Eq. (H.6.2).

$$d \geq 20\sqrt{\frac{10\gamma_0\gamma_{Q1}N_{\mathrm{p}}}{\pi f_{\mathrm{sd}}}} \tag{H.6.2}$$

where:

d—diameter of single steel tie rod (mm);

N_{p}—axial force in the tie rod (kN);

f_{sd}—design strength of the steel bar (MPa);this value may be adopted from the *Code for Design of Highway Reinforced Concrete and Prestressed Concrete Bridges and Culverts* (JTG D62);

γ_0—structural important coefficient that shall be adopted from Table H.1 of the *Specifications*;

γ_{Q1}—load partial factor of active earth pressure that shall be adopted from Table H.5 of the *Specifications*;

6 The area of the anchored bulkhead shall be calculated according to Eq. (H.6.3).

$$A = \frac{N_{\mathrm{p}}}{[p]} \tag{H.6.3}$$

where:

A—design area of anchored bulkhead (m^2);

[p]—allowable pullout force on a unit area of anchored bulkhead (kPa);this value shall be defined according to the data from an in-situ pullout test or from experience according to the in-situ conditions of engineering if the in-situ pullout test is not available.

7 The calculations for the retaining plate may be carried out according to the requirements of Clause H.5 of the *Specifications.*

8 The wall plate is calculated as a bending member supported by tie rods. If the wall plate is anchored by a tie rod, it may be calculated as a single point supported two-way cantilever slab and reinforcement need to be provided accordingly.

H.7 Reinforced earth retaining walls may be divided into two categories, with or without wall panels. The design and calculation should follow the specification of *Technical Sepcifications for Applications of Geosynthetics in Highway* (JTG/T D32) when the included angle between horizontal plane and enveloping geogrid reinforced slope surface without wall panels is less than 70°.When the included angel is greater than or equal to 70°,the design and calculation of reinforced earth retaining wall shall be conducted according to the following specifications:

1 Reinforced earth retaining wall design shall calculate internal stability and external stability. Carrying out the checking of external stability shall comply with the provisions of Clause H.2 in the *Specifications.* Settlement calculation shall be carried out for the reinforced earth body on soft subsoil. The global stability shall be checked for the reinforced earth body and roadbed to see if there may be a slide under the roadbed.

2 The buoyancy of the reinforced earth retaining wall that could be soaked in water shall be calculated according to the following requirements:

(1) The buoyancy at the low water level shall be used when reinforcement section has been designed.
(2) When the foundation bearing stress is checked, either buoyancy at the low water level or no buoyancy shall be used; when checking the sliding and overturn stabilities of the reinforced earth body, buoyancy at the design water level shall be used.
(3) For other conditions, the buoyancy at the most unfavorable water level shall be used.

3 The tension induced from the vehicle and pedestrian load shall be taken into account when designing the steel strap section. The steel strap pullout force caused by an additional load shall be disregarded in the calculation of strap anchorage length.

4 When checking the internal stability of reinforced earth body, the earth pressure coefficient shall be determined according to the following equation:

if $z_i \leq 6\ m$, then

$$K_i = K_i\left(1-\frac{z_i}{6}\right) + Ka\frac{z_i}{6} \tag{H.7.1}$$

If $z_i > 6\ m$ then

$$K_i = K_a \tag{H.7.2}$$

$$K_j = 1 - \sin \varphi \tag{H.7.3}$$

$$K_a = \tan^2\left(45° - \frac{\varphi}{2}\right) \tag{H.7.4}$$

where:

K_i—earth pressure coefficient at the depth of z_i in the reinforced earth body;

K_j—at-rest earth pressure coefficient;

K_a—active earth pressure coefficient;

z_i—distance between the node of the ith unit of steel strap and the top surface of the reinforced earth body;

φ—internal friction angle of fillers (°).

5 The horizontal earth pressure on wall plate $\sum \sigma_{Ei}$ shall be calculated according to the following equation:

$$\sum \sigma_{Ei} = \sigma_{zi} + \sigma_{ai} + \sigma_{bi} \tag{H.7.5}$$

where:

σ_{zi}—horizontal earth pressure on the wall plate at the depth of z_i that arises from the fill of reinforced earth (kPa);

σ_{ai}—horizontal earth pressure on the wall plate at the depth of z_i that is caused by vehicular (or pedestrian) load (kPa);

σ_{bi}—horizontal earth pressure on the wall plate at the depth of z_i that is caused by additional fill on top of the wall, which is converted to an equivalent uniform soil thickness (kPa).

6 The active region and stable region in the reinforced earth body may be divided by a simplified rupture plane as shown in Fig. H.3. The distance between the vertical part of the simplified rupture plane and the back surface of the wall plate b_H is equal to $0.3H$ and the angle between the inclining part of the simplified rupture plane and horizontal plane β is equal to $45° + \varphi/2$.

7 The additional load spread width may be determined according to a slope of 1:0.5 spreading down along the depth. The additional vertical earth pressure σ_{fi} (kPa) at the depth of z_i in the reinforced earth body is 0 and shall be calculated according to Eq. (H.7.6) when the spread line extends into the active region.

$$\sigma_{ji} = \gamma h_0 \frac{L_c}{L_{ci}} \tag{H.7.6}$$

where:

h_0—thickness of the equivalent uniform soil layer converted from vehicular or pedestrian (m) load;

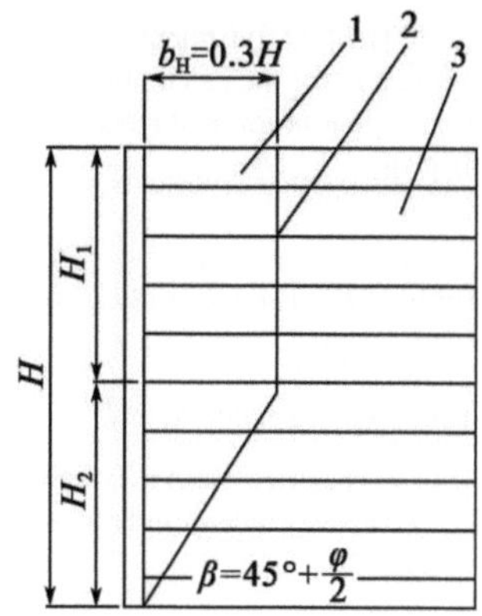

Fig. H.3 Illustration of the simplified rupture plane
1-Active region;
2-Simplified; 3-Stable region

L_c—loading width in calculation of the reinforced earth body (m), use the full substructure width;

L_{ci}—load spread width at the depth of z_i in the reinforced earth body (m);

γ—unit weight of the reinforced earth body (kN/m^3);for soaked retaining walls, it shall be taken separately for soil above and below the most unfavorable water level.

8 Under the permanent loads, the vertical earth pressure at the reinforcement location shall be calculated according to Eq. (H.7.7).

$$\sigma_i = \gamma z_i + \gamma h_1 \tag{H.7.7}$$

where:

σ_i—vertical earth pressure on the steel strap at the depth of z_i (kPa);

h_1—thickness of the converted equivalent soil layer top of the reinforced earth body (m).

9 The anti-pullout stability of the single node of steel strap shall be checked according to Eq. (H.7.8).

$$\begin{cases} \gamma_0 T_{i0} \le \frac{T_{pi}}{\gamma_{R1}} \\ T_{i0} = \gamma_{Q1} T_i \\ T_{pi} = 2f'\sigma_i b_i L_{ai} \\ T_i = (\sum \sigma_{Ei}) s_x s_y \end{cases} \tag{H.7.8}$$

In the calculation for anti-pullout resistance of the steel strap, the effects caused by the basic variable loads is be included.

Where:

γ_0—structure importance coefficient for Table H1 of the *Specifications*;

T_{i0}—design value of horizontal tensile force in the steel strap at the depth of z_i (kN);

T_i—horizontal tensile force in the steel strip at the depth of z_i;

$\sum \sigma_{Ei}$—horizontal earth pressure on the wall plate at the depth of z_i (kPa);

Table H.14 Adjustment coefficient for the calculation of anti-pullout force of steel strip γ_{R1}

Loading combination	I, II	III	Construction load
γ_{R1}	1.4	1.3	1.2

Table H.15 Pseudo-friction coefficient between fill and steel strip f'

Type of fill	Cohesive soil	Sandy soil	Granular soil
Pseudo-friction coefficient	0.25–0.40	0.35–0.45	0.40–0.50

Note 1. The coefficient may be increased by 0.1 if the steel strap is ribbed
2. The lower value of the ranges in this table is used when the height of retaining wall is larger than 12 m

γ_{Q1}—partial factor of the active earth pressure caused by the fill of the reinforced earth body and the additional fill on top of the retaining wall or the earth pressure caused by additional loads. It is taken from Table H.5 of the *Specifications*;

T_{pi}—pullout resistance of the effective length of steel strap loaded under the permanent gravity at the depth of z_i (kN);

γ_{R1}—adjustment coefficient for the calculation of the anti-pullout force of steel strap, which is taken from Table H.14;

s_x—horizontal distance between the adjacent nodes of steel strap (m);

s_y—vertical distance between the adjacent nodes of steel strap (m);

f'—pseudo-friction coefficient between the fill and steel strap; it is determined from a test or from Table H.15 if reliable test data is not available;

b_i—steel strap total width at node (m);

$L_{\alpha i}$—anchorage length of steel strap embedded in the stable region of the reinforced earth body (m).

10 The calculation for checking the tensile strength of the steel strap cross section shall meet the requirement of Eq. (H.7.9):

$$\gamma_0 T_{i0} \leq \frac{A f_k}{1000 \gamma_f \gamma_{R2}} \tag{H.7.9}$$

where:

A—effective net cross section area of the strap (mm^2);

f_k—standard valve of the steel strap of material strength as listed in Table H.16 (MPa);

γ_f—partial factor of tensile performance of the steel strap material, equal to 1.25 for all types of steel straps;

γ_{R2}—adjustment coefficient that may be taken from Table H.16 for the tension calculation of steel strap material.

11 The requirements for the effective net area A of the strap cross section shall be as follows:

Table H.16 Strength standard values f_k and adjustment coefficient γ_{R2} for the tension calculation of strap materials

Material type	f_k (MPa)	γ_{R2}
Q235 flat steel strap	240	1.0
Grade I reinforced concrete slab strap	240	1.05
Plastics-steel composite strap	Test fracture tension	1.55–2.0
Geotextile grid	Test fracture tension	1.8–2.5

Note 1. The large γ_{R2} values in Table H.16 for synthetic geotextile straps is used when the construction conditions are poor and the straps have large creep. The small γ_{R2} values may be used when the straps have small creep or when checking construction loading conditions
2. When a reinforced concrete strap is used, the reinforcement ratio shall be less than 2%
3. The elongation corresponding to the test fracture tension shall not be larger than 10%

(1) Flat steel strap: the net area is calculated based on the design thickness without the preset thickness for corrosion and excludes the holes for connections.
(2) Reinforced concrete strap; the tensile strength contributed by concrete is not considered; the effective net area of the reinforcement is the sum of all main rebar cross section areas, with the preset area deducted for corrosion.
(3) Plastics-steel composite strap: the dimensions of geotextile grids and polypropylene geotextile straps is provided by the manufacturers. The design cross section area and the ultimate strength is determined statistically with a guarantee rate of 98% through strict examination of the elongation and fracture strength.

12 The design calculation for wall panels shall be in accordance with the following requirements:

(1) The earth pressure is uniformly distributed on each panel.
(2) The panel is a simply-supported plate with cantilevers at both ends. The internal forces shall be calculated in vertical and horizontal directions, respectively.
(3) The connection between the strap and the wall panel should be properly strengthened.

13 When checking the anti-pullout stability of the entire wall, partial factors are all equal to 1.0, and shall meet the requirement of Eq. (H.7.10):

$$K_b = \frac{\sum T_{pi}}{\sum T_i} \tag{H.7.10}$$

where:

K_b—anti-pullout stability coefficient of the whole wall;
$\sum T_{pi}$—sum of friction forces of each strap layer;
$\sum T_i$—sum of tensile forces of each strap layer.

H.8 The calculation of ultimate limit state, checking calculation of serviceability limit state and structural requirements for the reinforced concrete members of pile-plank retaining wall shall comply not only with the *Specifications* but also with the requirements of the *Specifications for Design of Highway Reinforced Concrete and Prestressed Concrete Bridges and Culverts* (JTG D62) and *Specifications for Design of Ground Base and Foundation of Highway Bridges and Culverts* (JTG D63).

1 For the design calculations of reinforced concrete members of the pile-plank retaining wall, the values of structural important coefficient γ_0 shall be determined according to Sect. H.1 of the *Specifications* with the combination of action (or load) effects.

2 The more unfavorable action of either the landslide thrust or the earth pressure shall be selected as the load for the calculation of the pile-plank retaining wall that is located on the landslide substructure. The gravity of the pile itself may not be counted in the calculation.

3 The load width on the pile may be taken as the half the distance between the adjacent piles at each side; the load width on the plate may be taken as the plate calculation span.

4 The internal forces of the pile shall be calculated with the Foundation Coefficient Method in accordance with Clause 5.7.5 of the *Specifications.*

5 The ground stability and global stability shall be checked along the bedding landslide if the structural plane of the ground rock formation in front of the pile is tilting to the outside slope.

6 The precast reinforced concrete retaining plate may be assumed as a simply-supported slab supported by piles. The calculation span of the simply-supported plate is as follows:

$$\text{For circular pile } L = L_c - 1.5t \tag{H.8.1}$$

$$\text{For rectangular pile } L = L_0 + 1.5t \tag{H.8.2}$$

where:

L_c—center distance between adjacent piles with circular cross section (m);

L_0—clear distance between adjacent piles with rectangular cross section (m);

t—thickness of the retaining wall plate (m).

7 For a pile-plank retaining wall anchored by tie rods on the embankment, the occurrence of secondary stress induced by the fill settlement shall be avoided. The design of the tie rod shall be in accordance with the requirements in Sect. 5.5 of the *Specifications.*

Appendix J
Classification of Permafrost According to Highway Engineering

J.1 Permafrost shall be classified in accordance with volume ice content and according to Table J.1. Permafrost can be divided into low ice frozen soil, icy frozen soil, rich ice frozen soil, saturated ice frozen soil and ice layer containing soil. Among them, rich ice frozen soil, saturated ice frozen soil and ice layer containing soil are collectively called frozen soil with high ice content.

J.2 According to the annual average ground temperatures, permafrost may be divided into low temperature frozen soil (annual average ground temperatures ≤ 1.5 °C) and high temperature frozen soil (annual average ground temperatures > 1.5 °C).Stability types of frozen soil areas may be classified according to Tables J.2 and J.3.

W. Wu, *Specifications for Design of Highway Substructure*, JTG Standards and Specifications, https://doi.org/10.1007/978-981-96-2313-6

Table J.1 Classification of the permafrost and class of thaw settlement

Permafrost type	Soil type		Total moisture content ω_n (%)	Volume ice content i	Average thaw settlement factor δ_0 (%)	Thaw settlement class	Thaw settlement type
Low ice frozen soil	Coarse-grained soils	Silty and clay content ≤15%	<10	<0.1	≤1	I	No thaw settlement
		Silty and clay content >15%	<12				
	Fine sand, silty sand		<14				
	Silt		<17				
	Cohesive soil		$<\omega_p$				
Icy frozen soil	Coarse-grained soils	Silty and clay content ≤15%	10–15	0.1–0.2	$1 < \delta_0 \leq 3$	II	Weak thaw settlement
		Silty and clay content >15%	12–15				
	Fine sand, silty sand		14–18				
	Silt		17–21				
	Cohesive soil		$\omega_p < \omega_n < \omega_p + 4$				
Rich ice frozen soil	Coarse-grained soils	Silty and clay content ≤15%	15–25	0.2–0.3	$3 < \delta_0 \leq 10$	III	Thaw settlement
		Silty and clay content >15%	15–25				
	Fine sand, silty sand		18–28				
	Silt		21–32				
	Cohesive soil		$\omega_p + 4 < \omega_n < \omega_p + 15$				

(continued)

Table J.1 (continued)

Permafrost type	Soil type		Total moisture content ω_n (%)	Volume ice content i	Average thaw settlement factor δ_0 (%)	Thaw settlement class	Thaw settlement type
Saturated ice frozen soil	Coarse-grained soils	Silty and clay content ≤15%	25–44	0.3–0.5	$10 < \delta_0 \leq 25$	IV	Strong thaw settlement
		Silty and clay content >15%	25–44				
	Fine sand, silty sand		28–44				
	Silt		32–44				
	Cohesive soil		$\omega_p + 15 < \omega_n < \omega_p + 35$				
Ice layer containing soil	Coarse-grained soils	Silty and clay content ≤15%	>44	>0.5	>25	V	Thaw collapse
		Silty and clay content >15%					
	Fine sand, silty sand						
	Silt						
	Cohesive soil		$> \omega_p + 35$				

Note 1. The Coarse-grained soils include crushed (gravel) stone, gravel sand, coarse sand and medium sand
2. Total moisture content includes ice and unfrozen water
3. ω_p is the plastic limit moisture content
4. Frozen saline soil, peatized frozen soil, humic soil and high plastic clay soil are not included

Table J.2 Classification of stability types of frozen soil areas

Annual average ground temperatures(°C)	≥−0.5	−0.5–≤ −1.0	−1.0–≤ −2.0	<−2.0
Stability types of frozen soil areas	High temperature extremely unstable frozen soil area	High temperature unstable frozen soil area	Low temperature basically stable frozen soil area	Low temperature stable frozen soil area

Table J.3 Classification of stability types of various frozen soil areas

Permafrost type	Low ice frozen soil	Icy frozen soil		Rich ice frozen soil		Saturated ice frozen soil			Ice layer containing soil		
Annual average ground temperatures (°C)	No consideration	0–1.0	<−1.0	0–1.5	<−1.5	0–1.0	−1.0–−2.0	<−2.0	0–1.0	−1.0–2.0	<−2.0
Stability types	Stable	Basically stable	Stable	Basically stable	Stable	Unstable	Basically stable	Stable	Unstable	Basically stable	Stable

Wording Explanation for the *Specifications*

1 The strictness in execution of the *Specifications* is expressed by using the wording as follows:

(1) MUST—A very restrict requirement in any circumstances.
(2) SHALL—A mandatory requirement in normal circumstances.
(3) SHOULD—An advisory requirement.
(4) MAY—A permissive condition. No requirement is intended.

2 Expressions used for reference to standards are explained as follows:

The standards for which a year is added to the standard number shall be the specific versions to be used. Otherwise they shall be the latest available versions.

W. Wu, *Specifications for Design of Highway Substructure*, JTG Standards and Specifications, https://doi.org/10.1007/978-981-96-2313-6

Background to Provisions

1. General Provisions

1.3 This section specifies the basic requirements for substructures performance. The substructure consists of subgrade and substructure facilities. The subgrade refers to the range under the pavement structural layer; substructure facilities refer to the necessary ancillary engineering facilities used to ensure the roadway stability. It includes drainage facilities, protection facilities and retaining wall.

The roadbed structure is banded geotechnical structure under the pavement layer, and is the basis of the pavement, which needs to bear the affects from the car load and the role of various natural factors, and is consistent with the English 'Substructure'. Under the influence of various environmental factors (wind, rain, snow, temperature, water flow, earthquake, etc.) and the impact of vehicle load, the strength and rigidity of the roadbed will be attenuated, which will affect the load carrying capacity of the substructure, resulting in settlement deformation and slip damage. Therefore, in the process of design, it need to take the environmental factors and the impact of vehicle load on the long-term performance of the substructure into account, so that the substructure has enough strength, stability and durability to prevent substructure disease and to ensure highway operation safety.

1.4 Geological data is the basis of substructure design. In the process of roadbed construction, especially in mountainous areas, due to the complex and flexible geological situation, subject to a variety of factors, accuracy and reliability of geological survey data is not high, so it is easy to find that the design and the actual situation does not match after substructure construction excavation. Or in complex geological conditions, the unbefitting construction methods and improper process results in new diseases and so on. Therefore, this part emphasizes the need to strengthen the geological exploration test work and to identify the geological conditions along the highway, lay a solid foundation for the highway design.

1.5 The height of substructure is a comprehensive technical and economic index in highway design which directly affects the function, quality, project cost, area and surrounding landscape of highway. The high embankment and road bed (substructure

W. Wu, *Specifications for Design of Highway Substructure*, JTG Standards and Specifications, https://doi.org/10.1007/978-981-96-2313-6

work area) are in dry and wet states or dry states, and the long-term performance of the substructure is relatively stable, but it is more expensive and the project cost is higher, which is prone to settlement deformation and slope stability. Low embankment can save the land, have small impact on the environment, but the climate, environment and groundwater will have a significant impact on the roadbed (roadbed work area) performance, resulting in lack of load capacity of the road bed, causing the deformation of the road damage. So the low embankment needs high demand on the roadbed filler and underground drainage measures. In addition, in the villages, local roads, navigable river-intensive areas, low embankment program need to be merged channel, add auxiliary roads, make cross-section and so on. The construction scale of supporting project is huge, low embankment program advantage is not significant.

Therefore, according to the project where the terrain, geology, hydrology and other natural conditions, as well as distribution characteristics of villages, towns, waterways, and highway network, design need to demonstrate with comprehensive comparison about different substructure height and filling program and make the adaptation to local conditions. Under the premise of satisfying the requirements of road function and the requirement of roadbed performance, it is reasonable to determine the height of substructure and filling scheme, also make the comprehensive design of roadbed pavement.

3. General Substructure

3.1 General requirements

3.1.2 Mountain highway terrain and geology is more complex, it is restricted by the highway route plane and vertical technical indicators standard. So there are so many high-filling and deep-cutting substructures. From highway construction situations, when the soil or soil like (full-strong weathered layer) cutting slope height is greater than 20 m, rock cutting slope height is greater than 30 m, the stability is poor and there are a large amount of unstable high slope. In order to solve the high slope stability, it takes some reinforcement projects like a large number of prestressed-anchor cables, anti-slide pile or anti-slide wall. Not only the project cost is high, but also affect the surrounding environment.

High embankment slope height is 20–30 m, some engineering embankment slope is up to 40–50 m, which is often accompanied by steep slope embankment. There are slope stability and embankment uneven deformation problems, and also take up a lot of land. From the overall socio-economic benefits, high embankment is often not the best project.

In summary, the design of mountain substructure requires appropriate restrictions on the height of high embankments and deep cuts. It is necessary to increase the intensity of the route scheme by setting up the settlement and stabilization of the high embankment and deep cuttings in the preliminary design stage. In the comparison of the program, it is necessary to consider the construction period of technical complexity, project cost, construction methods, but also to consider the road during the operation due to environmental impact and possible diseases and maintenance costs, as well as socio-economic environmental benefits. In the case of almost the same investment projects, it is preferred to choose bridge and tunnel engineering program.

3.1.3 Submerged substructures include long-term submerged substructure and temporary rainwater substructure during rainy season. Long-term submerged substructure refers to the embankment on one side or both sides of the road, which is flooded long-term, including the substructures adjacent ponds, rivers, reservoirs and so on. The temporary flood dike in the rainy season is the substructure which is affected by floods on one side or both sides of the substructure, including the floodplain, flood detention area, flood diversion area and so on. In order to ensure the safety and stability of the substructure and meet the functional requirements of the flood relief channel, the *Specifications* specifies the flood frequency of the substructure along the river and flooded according to the relevant provisions of the existing *Highway Engineering Technical Standard* (JTG B01).

3.1.5 On the highway construction, the proportion of earth and stone in substructure is high, take and spoil occupied the more land. In order to save investment, less farmland, in the design it makes full use of substructure excavation materials, and it is necessary to rational allocation of earth and stone and reasonable set aside the soil abandoning area. At the same time, for those who cannot meet the strength requirements of the excavation material, it should not easily abandon. Combined the soil treatment and inspection process and construction technology, make a technical and economic comparison between remote transport and soil extraction plan and soil improvement treatment program, then choose the better one. When the cost of almost the same circumstances, as far as possible choose the cutting soil improving treatment program to reduce the number of taking and abandoned soil in substructure and save the land.

3.1.8 High embankment, steep embankment and deep cutting, due to mountainous terrain and geological complexity, geological survey data cannot fully reflect the actual situation, construction methods and processes also affect the high slope stability and embankment dense state, the design is also difficult to simulate the actual construction status. By dynamic monitoring of construction, it can not only validate and improve the design to ensure stable roadbed, but also estimate the amount of the settlement after the embankment completed to determine a reasonable time for paving road surface, and to effectively control of high embankment uneven subsidence. Therefore, it is necessary to adopt dynamic design method for highway, high-side embankment, steep-slope embankment and excavation high slope substructure.

Dynamic design is the basis principle of substructure design. The dynamic design improves itself based on the feedback information and monitoring data in the construction. It is an objective and realistic, accurate and safe design method. It is suitable for the construction of the substructure and is an extension of the construction drawing design. Dynamic design should take complete construction design drawing as a basis, not the name of 'dynamic design' under the banner of 'side construction, side design'. At the same time, to correctly deal with the dynamic design and change the design. The dynamic design is to perfect and optimize the original design rather than change major design of the engineering program.

3.2 Roadbed

3.2.1 The original *Specifiations* (JTG D30—2004), did not explicitly put forward the terminology and stratigraphic structure of the roadbed structure, but followed the division of the substructure. Substructure usually refers to the depth of the substructure work area below the pavement structure, the additional stress generated by the car load is relatively significant, and is closely related to the pavement structure response, consistent with the English 'Substructure'. This revision, in line with international standards, has been proposed this section 'roadbed' to 'substructure'. In the process of soliciting opinions and review, we believe that 'substructure' is easy to cause ambiguity. Therefore, this revision still follows the traditional concept of substructure layer-upper bed and lower road bed to approximate the definition of highway substructure, and clear its technical requirements, design indicators and related technical measures.

Substructure and the scope of the road is basically the same, are based on the depth of the roadbed work area and the original *Specifications* (JTG D30—2004) road bed range is 0.8 m. In recent years, with the continuous changes of axle load spectrum and its significant effect on roadbed performance, the research and discussion on the impact of vehicle load on road surface performance and the depth of roadbed work area are very active. Although the standard and the analysis result of the substructure work area are different, it is generally believed that the depth of the working area of highway substructure in China is obviously larger than 0.8 m. AASHTO (1993) clarifies that the impact of vehicle loads on substructure is 1.5 m. Tongji University of China make the numerical analysis on more than 40 kinds of typical asphalt pavement structure of the substructure work area depth, take 95% of the cumulative frequency corresponding to the value shown in Table 3.1; through the test road measured the depth of the working area of the asphalt pavement is 0.9–1.0 m($\sigma_z/\sigma_c \leq 0.2$)or 1.3–1.5 m ($\sigma_z/\sigma_c \leq 0.1$) under the single-axis dual-wheel 100 kN standard axle conditions, while the depth of the asphalt pavement 1.6–2.0 m ($\sigma_z/\sigma_c \leq 0.1$) under the 130 kN overload conditions. The numerical analysis is consistent with the measured results, and it is proved that the depth of the roadbed work area in our country is obviously higher than that of the road bed (0.8 m).

When the *Specifications* revised, according to the actual situation of research and engineering, it adjusts the range of the road bed. Light, medium and heavy traffic roadbeds thickness are 0.8 m, heavy, heavy traffic highway roadbeds thickness are 1.20 m. For special axle load roads, it is necessary to determine the thickness of the road bed by calculating the depth of the substructure work area.

For the road bed division, from the project economy considerations, the up roadbed is still taken as 0–0.3 m; down roadbed is in accordance with the traffic load level division. For light, medium and heavy traffic road it is still 0.3–0.8 m. For special heavy, heavy traffic road the down roadbed is revised to 0.3–1.2 m.

Special road axle load means is coal or other large building supplies based truck roads, according to the actual need, after investigation it demonstrated individually

Table 3.1 Analysis results about substructure work area depth

Axle type and single axle load	Different identification criteria corresponding to the substructure work area depth (m)		
	$\sigma_z/\sigma_c \leq 0.1$	$\sigma_z/\sigma_c \leq 0.2$	$\sigma_z/\sigma_{zo} \leq 0.25$
Single-axis dual-wheel 100 kN	1.3	0.9	1.9
Three-axis dual-wheel 130 kN	2.4	1.6	3.0

Note 1. σ_z is the vertical stress that is transmitted to the substructure through the pavement structure
2. σ_c is the vertical stress caused by overburden structure
3. σ_{zo} is the vertical stress for the vehicle load through the pavement structure to the top of the substructure

selected axle load calculation parameter, is calculated to determine the thickness of the road bed.

3.2.2 In the analysis and design of pavement structure, the index of roadbed structure is the rebound modulus of the top of the substructure (ie, the top of the road bed). CBR is an indicator of the water stability of the material and the ability to resist local buckling, while the compaction degree is the construction control index, both of which are not substructure design indicators (or parameters). However, the design value of the rebound modulus of the roadbed must be achieved through the reasonable selection of the packing and the effective control of the compaction degree. The engineering practice has proved this well. Therefore, this revision retains the relevant provisions of the original *Specifications* (JTG D30—2004) using the minimum strength of substructure packing (CBR) and roadbed compaction, and changes the 'minimum strength' to 'minimum load ratio'.

For the CBR of filler, on the one hand, the minimum value of most countries is slightly higher than that of our country. On the other hand, the vast majority of highway substructure is long-term in the unsaturated state, especially in China's dry and semi-dry areas the equilibrium moisture is below good water content, and thus there is a clear irrationality in the current CBR test method saturated water 4 d test conditions. Since considering these two aspects, this revision still maintains the minimum CBR standard for the original *Specifications* (JTG D30—2004).

On the degree of compaction, the original *Specifications* (JTG D30—2004) has been raised by 1–3 percentage points on the basis of JTG 013—95, which has played an important role in ensuring the performance of substructure. But relatively speaking, this standard is not high. For example, AASHTO (1993) requires a minimum compaction degree of more than 95% of the recommended embankment and roadbed structure, except for special soils. For some soil groups even require 100%. Taking into account the further increase in the degree of compaction in the actual operation of the more difficult, not economic, so this revision maintain the original *Specifications* (JTG D30—2004) of the roadbed compaction requirements.

3.2.4 On the design indicators of subgrade structural performance, the domestic has made the system research in the past decade. The substructure soil is a nonlinear elastoplastic material, and the elastic modulus value of the stress–strain relationship is stress-dependent and varies with its humidity and density. The current design specification uses the static load plate method to measure the rebound modulus of the substructure on the top surface, or use the Beckman beam to measure the surface backlash after bending. The reference value determined by the look-up table method in the *Code for Design of Highway Asphalt Pavement* (JTG D50—2006) is based on the value of the national survey and calculation analysis in the 1970s. Whether it is the test method, the parameter index or the reference value, they cannot accurately reflect the basic mechanical properties of the material under the load.

The permanent deformation of the substructure is an integral part of the asphalt pavement rut. The methods of estimating the permanent deformation of the substructure are the hierarchical strain sum method and the control method based on the stability theory. The accumulation of permanent deformation of the substructure is related to the level of repetitive stress. When the stress level is high, the cumulative rate of permanent deformation will increase with the increase of the number of times, which will lead to excessive deformation or failure after the action of the number of times. When the stress level is low, the cumulative rate of permanent deformation increases with the increase of the number of times and gradually slows down, and tends to stabilize the state. According to the theory of stability, the stress or strain level of the roadbed is controlled so that the permanent deformation accumulated can eventually reach the equilibrium (steady) state, and the permanent deformation of the roadbed can be controlled accordingly. It results in damage due to excessive permanent deformation of the roadbed. Compared with the stratified strain method, the application of allowable stress or strain index to control the permanent deformation of the roadbed is easier to implement.

The revision is based on the Ministry of Transport in the western transport construction technology project 'asphalt pavement design indicators and parameters of' 'cement concrete pavement substructure performance indicators and parameters' 'based on multi-index asphalt pavement structure design method' 'embankment reasonable height of the study' 'the research on the performance and design of highway subgrade structure and other related indexes'. This paper draws on the experience of mechanistic experience of pavement structure design, and uses the dynamic rebound modulus of the top surface of the roadbed as the design index. Face vertical pressure strain for the calculation of indicators. Revised *Highway Asphalt Pavement Design Code* is also the use of dynamic resilience modulus as a design index, vertical pressure strain as a calculation index.

The design control standard of the dynamic rebound modulus of the top of the road bed is that substructures in the balance humidity state should not be lower than the requirements from current *Specifications for Design of Highway Asphalt Pavement* (JTG D50) and *Specifications for Design of Highway Cement Concrete Pavement* (JTG D40).

Clause 4.2.2 of the *Design Code for Highway Cement Concrete Pavement* (JTG D40—2011) stipulates that the integrated backfill modulus value on the top of the roadbed shall not be less than 40 MPa in the light traffic load level, and shall not be less than 60 MPa in the medium or heavy traffic load level, extra heavy or extreme heavy traffic load level shall not be less than 80 MPa.

Being revised *Highway Asphalt Pavement Design Code* (JTG D50) (Manuscript) requires that the backfill modulus value on the top of the roadbed is not less than 40 MPa in the light traffic load level, and is not less than 60 MPa in the medium or heavy traffic load level, extra heavy traffic load level is not less than 90 MPa, extreme heavy traffic load level is not less than 120 MPa.

Substructure balance humidity refers to the road after completion of traffic, the substructure in the groundwater, rainfall, evaporation, freezing and melting and other factors, its humidity reaches a relatively stable equilibrium state, humidity is called balance humidity.

3.2.5 The current design specification only selects the rebound modulus value according to the substructure humidity condition of the most unfavorable period. The design state is not very clear when the seasonal variation of the roadbed humidity is not fully considered in the design. It is not possible to establish the quantitative relationship between the humidity, the rebound modulus and the substructure humidity and the rebound modulus during the roadbed construction. But also did not fully consider whether the substructure rebound design value can meet the traffic load requirements.

The results of the existing substructure survey and indoor test show that under the influence of natural environment (rainfall, evaporation, freezing, melting, etc.) and groundwater, the new water transfer and humidity distribution will be generated in the substructure. Water content increased by 2–10 percentage points compared with that at construction time. The LTPP 137 substructure (78 aggregates and 59 cohesive soils) surveyed in the United States, the moisture content increased with the increase of humidity, roadbed strength and resilient modulus The same time, the substructure in the wet and dry cycle, freeze–thaw cycle process, will also damage the substructure soil structure, making the substructure soil strength and resilience modulus attenuation. According to the research results, the variation law of humidity, strength and elastic modulus of substructure soil under the conditions of humidity, strength, rebound modulus and equilibrium state of substructure soil are obtained.

According to the research results, the *Specifications* revised that with the best moisture content and the maximum dry density of the roadbed humidity as the standard state, it take the substructure rebound modulus as a design value, and take full account of humidity changes, wet and dry cycle or freeze–thaw cycle in the effect of the rebound modulus on the substructure which make the balance humidity state not lower than that of the pavement design specifications. That is, the design value of the elastic modulus of the homogeneous soil substructure of the new highway is determined according to the Eq. (3.2.5.1) and the requirements of Eq. (3.2.5.2) are satisfied.

The recovery modulus of the substructure under the dry and wet cycle or the freeze–thaw cycle is in the range of 0.7 to 0.95. Dry and dry cycle conditions refer to non-frozen areas, the reduction factor is closely related to the substructure humidity state, soil type and possible water loss rate, and the wetting and wetting state is fine. When the possible water loss rate is small, the reduction coefficient takes a large value; the coarse grain (such as gravel) reduction coefficient takes a large value. The freeze–thaw cycle condition is the seasonal frozen soil area, and its reduction coefficient is closely related to the freezing temperature, the roadbed humidity state and the soil type. The light-frozen area fine-grained soil substructure, the reduction coefficient is larger; area of wet, in the wet state of fine-grained soil substructure, the reduction factor to take a small value; coarse grain (such as gravel) reduction coefficient to take a large value.

For the multi-layer and different types of soil substructure, the multilayer elastic layered foundation theory is used to convert the multi-layer structure into equivalent single-layer structure according to the principle of deflection equivalent, and then calculate the equivalent modulus of the top of the road bed.

3.2.6 The modulus of elasticity of the substructure soil under the standard condition of the new highway is determined by repeated loading triaxial compression test. Taking into account a lot of design units have not yet equipped with soil dynamic three-axis instrument, the *Specifications* revised as a transition, in the early large number of tests on the basis of the Appendix B is also given the dynamic rebound modulus reference value, and the elastic modulus. The relationship between the design may be used to look up the table or change the method to determine the substructure soil resilience modulus value.

Experiment Study related factors, stress analysis condition, physical condition and the nature and composition of the resilient modulus values, rebound modulus establish constitutive soil model reflects nonlinear behavior and pellets. Thus, on the basis of physical property values of test water content w, substructure dry density p_d, plasticity index I_P, the fine content $P_{0.075}$, using a three constitutive model parameters of Eq. (3.1) may be determined substructure soil rebound modulus value.

$$M_R = k_1 p_a \left(\frac{\theta}{p_a}\right) k_2 \left(\frac{\tau_{oct}}{p_a} + 1\right) k_3 \quad (3.1)$$

where:

M_R—modulus of substructure resilience (MPa);

p_a—the absolute value of atmospheric pressure, usually 100 kPa;

θ—the body stress (the first stress invariant) is the sum of the three principal stresses, $\theta = \sigma_1 + \sigma_2 + \sigma_3$ for intermediate principal stress (kPa);

τ_{oct}—eight body shear stress (kPa);

$$\tau_{oct} = \sqrt{(\sigma_1 - \sigma_2)^2 + (\sigma_2 - \sigma_3)^2 + (\sigma_3 - \sigma_1)^2/3}$$

k_1, k_2, k_3—model parameters; substructure for fine-grained soil, moisture content may be w is(%)substructure dry density p_d (g/cm^3), I_P plasticity index(%),fine content $P_{0.075}$(%)physical properties such as index, according to Eq. (3.2) empirical estimates of three parameters.

$$\begin{aligned} k_1 &= -0.096{,}0w + 0.392{,}9p_d + 0.014{,}2I_P + 0.010{,}9P_{0.075} + 1.010{,}0 \\ k_2 &= -0.000{,}5w - 0.006{,}9I_P - 0.002{,}6P_{0.075} + 0.698{,}4 \\ k_3 &= -0.218{,}0w - 3.025{,}3p_d - 0.032{,}3I_P + 7.147{,}4 \end{aligned} \quad (3.2)$$

According to the research results, the variation range of the equivalent stress level in the substructure of different traffic load grades is small. When determining the substructure equivalent elastic modulus, the equivalent stress level may be obtained by the body stress of 0 to 70 kPa and the octahedral shear stress of 13 kPa.

3.2.7 The current specification divides the roadbed into four types according to its wet and dry state: wet, wet, medium and dry, and the average consistency of 80 cm depth below the top of the bed is used as the index of the roadbed humidity. The wet and dry conditions of the four types of roadbeds are classified according to the critical height of the substructure, and the standard of the classification follows the national survey results from the late 1950s to the early 1960s.

The roadbed humidity conditions are affected by various factors such as atmospheric precipitation and evaporation, groundwater, temperature and pavement structure and water permeability. Many observations show that the humidity change of the substructure is gradually approaching a certain equilibrium humidity condition within 2 to 3 years after the completion of the road surface.

The specifications revised that, according to the roadbed of the humidity source, the substructure can be divided into three types of equilibrium humidity:

(1) Groundwater control type of roadbed-high groundwater level, roadbed work area in the groundwater capillary wetland area, the roadbed balance humidity controlled by the groundwater level movements. This substructure humidity state is defined as a wet state.
(2) The climatic factors control the substructure that the groundwater level is very low, the substructure work area is above the groundwater capillary wetting area, and the substructure equilibrium humidity is controlled by the climatic factors. This substructure humidity state is defined as dry.
(3) Substructure affected by groundwater and climatic factors-groundwater level is higher, the lower part of the substructure work area is in the area of groundwater capillary wetland, and its upper part is affected by climatic factors. The roadbed equilibrium humidity is affected by groundwater and climate two aspects of the impact. This substructure humidity state is defined as the medium wet state.

The current specification uses the consistency to characterize the humidity of the substructure, on the one hand cannot reflect the non-cohesive soil moisture state, on the other hand, the moisture content of moisture alone, it is difficult to accurately reflect its impact on the elastic modulus. This revision uses saturation to characterize the humidity of the substructure soil. The saturation of the soil reflects both the moisture content and the influence of the density.

Saturation is determined by:

$$S_r = \frac{\omega_V}{1 - \frac{\gamma_s}{G_s \gamma_w}} \quad or \quad S_r = \frac{\omega}{\frac{\gamma_w}{\gamma_s} - \frac{1}{G_s}} \tag{3.3}$$

$$\omega_V = \omega \frac{\gamma_s}{\gamma_w} \tag{3.4}$$

where:

S_r—saturation (%);
ω_v—volume moisture content (%);
ω—mass moisture content (%);
γ_s, γ_w—soil dry density and water density (kg/m^3);
G_s—relative density of soil.

The prediction of substructure equilibrium moisture is mainly based on the soil water characteristic curve of unsaturated soil mechanics (saturation or water content, matric suction relationship curve). By groundwater control, the groundwater level model is used to predict the substructure matric suction. By climatic factors control, the TMI model is used to predict the substructure matric suction, and the TMI value shall be checked according to Eq. (3.5). The TMI value in different natural zoning is calculated according to the more than 400 meteorological stations and the meteorological data after merging. In view of the relatively complex theoretical calculation, the look-up table method of Appendix C is given.

$$\mathrm{TMI}_y = \frac{100R_y - 60\mathrm{DF}_y}{\mathrm{PE}_y} \tag{3.5}$$

where:
R_y—water runoff (cm) in y;
DF_y—water deficit (cm) in y;
PE_y—potential evaporation (cm) in y.

3.3 Filling Embankment

3.3.1, 3.3.2 Minimum height of fill substructure is determined in the current specification, the main consideration of substructure design of flood and roadbed in the centralized wet state of critical height, not yet fully considering the influence of groundwater on the properties of substructure work area, and the seasonal frozen region substructure frost depth and freezing water rising influence on substructure. The high level of underground water section of low embankment of roadbed, groundwater (substructure work area) performance based soil strength resulting from the attenuation caused by the bearing capacity of substructure work area is insufficient, under vehicle load, substructure will produce large plastic deformation, deformation and damage caused by pavement.

The revised standard, based on summarizing the experience of engineering and scientific research, proposed to determine the height of embankment is considered when the design flood level, in the wet state of substructure critical height, depth, substructure work area of substructure freezing depth and other factors, and determine the method of calculating the height of embankment type [see Eq. (3.3.2)].

3.3.4 Special arid area refers to the annual precipitation is very small, generally not more than 200–250 mm, strong evaporation area, such as desert, Gobi; special humid area refers to the annual rainfall, generally more than 1,000 mm, the rainy season lasts for several months, and the soil is too wet state of clay soil. In the special drought, special humid area, the roadbed compaction is very difficult, the provisions of this specification 'in the premise of the strength and modulus of substructure, through the test, the compaction standard can be reduced by 1–2 percentage points'.

3.3.5 According to the existing engineering experience and research results, the steepest good foundation embankment stability requirements under typical soil embankment slope was given based on. In the design, according to the geological conditions and the embankment height, fillers, protection and land type, flexible design of embankment slope rate, the conditions, in addition to Gobi, farmland, desert, forest officials etc., to slow down the rate of embankment slope, road base and the surrounding environment integration and try to provide appropriate opportunities for emergency vehicles out of control.

3.3.7 There is often a phenomenon of vehicle jumping between the embankment and the bridge, the culvert and the passageway. The main reason is that the compaction of the embankment is not enough. In order to eliminate this jumping phenomenon, it is an effective engineering measure to set up transition section between the embankment and abutment, transverse structure (culvert, channel) and so on. For transition section lengths, the JTG D30—2004 is used (2–3) *H*. According to the actual engineering situation, this revision has increased the length of transition section for the convenience of construction of rolling machinery.

3.3.9–3.3.11 Shoulder stone roadbed, roadbed, roadbed foot has economical characteristics, good effect, and is widely used in highway. When designing, we should give full consideration to the masonry foundation, the foundation conditions and the types of stone used, and require the slope foundation to be stable and have good bearing capacity. The masonry stone should be made of hard rock, and the soft rock cannot be used. Masonry substructure is only applicable to grade three or four highway.

3.4 Cutting substructure

3.4.1 Table 3.8 cutting slope rate according to that the soil is homogeneous, no bad geological phenomena and groundwater conditions, meet the requirements of road slope soil excavation slope stability the slope rate of stem. The design shall be according to the site, climate, geology and other natural factors such as the results of the survey, combining with geological conditions, slope height and form of protection, flexible design of soil excavation slope ratios, when conditions may be appropriate to slow slope rate.

3.4.2 Stability analysis and design of the rock slope is more complex, affected by the lithology, slope height and construction method and other factors, still depends largely on the rock structure, structural plane occurrence and degree of weathering. How to correctly judge and balance the influence degree of various factors on slope stability, and make more accurate and reliable quantitative analysis and slope stability evaluation, there is no unified and perfect method at present.

The stability of rock slope is mainly controlled by the rock mass structure, according to the *Technical Code for Building Slope Engineering* (CB 50,330—2002) Table 12.2.2, combined with the characteristics and experience of highway slope, the slope stability according to the status of different rock types, are presented in Table 3.9 of slope height not more than 30 m and no camber of weak structural plane of rock slope, rock slope according to the classification of Appendix E to determine. Since our country is vast in territory, topography, geology and climate change are great, we should pay attention to study and accumulate the slope stability rate of various rock mass types at different slopes, for reference to the next revision.

The stability of rock cutting slopes is greatly affected by construction technology and construction methods. For the hard rock excavation slope, conventional method, impacted by blasting and earthquake, will make the rock broken, loose, easy to produce during the operation of block, rock fall or landslide collapse disease. Therefore, it is not necessary to make use of the characteristics of natural rock mass in design, so the rock mass characteristics after blasting should be considered.

The engineering practice proves that the use of smooth blasting and presplitting blasting and millisecond blasting technology can improve the quality of cutting slope, slope damage to minimize excavation, stability, slope rock mass formed after the formation of beautiful construction, worth popularizing.

3.4.4, 3.4.5 Water is one of the important factors affecting the stability of cutting slope, rainfall infiltration and groundwater infiltration will make the discontinuities in the saturated state, reduce the soil shear strength, will cause the slope stability problem caused by deformation. Therefore, the design of slope water control and drainage is very important. When designing, the necessary drainage facilities should be set up according to the distribution of surface water and groundwater and the local conditions.

3.5 Boundary region treatment between the excavation and fill substructure

3.5.1, 3.5.2 Semi filled and semi excavated substructure is widely distributed in mountain highway, and the roadbed of the substructure is often different from settlement, deformation and destruction. The main reason is the difference between the material properties and the dense state of the filling and excavation, as well as the groundwater.

To reduce the differential settlement of substructure, the first from the fill material design staff, fill the selected packing need to match the characteristics of rock and soil excavation area, the conditions, the preferred good water permeability of coarse grained soil filling, can reduce the differential settlement, and provides a path for the excavation of groundwater discharge, avoid the filler permeability sealing excavation zone of groundwater, causing diseases of substructure.

According to the experience of highway construction, after filling the substructure to reach the required compaction requirements, measures such as impact, rolling or heavy tamping can be adopted to effectively reduce the differential settlement of the excavation and excavation joint. But the surface soil loose and easy to grind punch after roll and impact, roller compaction or for embankment of roadbed bottom to reinforce or improve pressure.

3.5.3, 3.5.4 The stability of the semi filled and semi excavated substructure on the steep slope is poor, some of which are designed to use the multi-layer geogrid on the steep slope to improve the stability of the semi filled and semi excavated substructure. From the practical effect, the effect is not significant, so the provisions when excavating and filling between substructure stability is not enough, according to the geological conditions, to improve the basal conditions of the foundation, setting mortar masonry protection and retaining wall retaining engineering.

3.5.5 The deformation and failure of substructure in semi filled and semi excavated substructure are mostly caused by ground water. When the underground drainage system of the substructure is not perfect or the discharge capacity is insufficient, the filling and excavation combined with the substructure will cause water accumulation and soften the substructure and cause the deformation and failure of the substructure and pavement. According to the investigation, the underground water drainage system has not been found, so it is very important to set up a perfect underground drainage system for the semi filled and semi excavated substructure.

3.5.6 Longitudinal roadbed fill with the department, especially between rock fill embankment and excavation section, because of the difference of excavation section and filling material properties, and the underground water seepage, deformation and failure often cause the deformation of roadbed disease. In the longitudinal section of the substructure, the transition section of the joint is provided with the sand gravel and gravel soil with good water permeability, so that the substructure diseases can be well prevented.

3.6 High embankment and steep slope embankment

3.6.1, 3.6.2 The original *Specifications* (JTG D30—2004) specifies that the embankment with slope height exceeding 20 m is a high slope embankment, and that the slope of the transverse slope is more than 1:2.5 is a steep slope embankment. In this standard revision, it is called 'high embankment' and 'steep slope embankment'. According to the investigation situation, there are still many problems that appear to be concerned with stability, which is still the type of substructure. Therefore, the standard emphasizes on individual geological survey design, integrated design and dynamic design.

Bad geology means landslide, collapse, mud rock flow and so on. Special rock soil refers to soft soil, red clay, high liquid limit soil, swelling soil and loess (collapsible loess). The embankment unfavorable geology, special geotechnical section for construction site design in accordance with the relevant provisions of chapter seventh.

3.6.5 The adoption of gentle slope rate is beneficial to slope stability, but the land cover is increased. Therefore, the provisions shall be comprehensively determined according to the topography and engineering geological conditions, the slope height and stability of the substructure, combined with the economic and environmental factors.

3.6.6 The slope of high fill substructure is usually wide and affected by unfavorable factors such as rain. China Merchants Chongqing communications research and Design Institute Co. Ltd. was studied on the seepage water in the form of high embankment slope on the rainwater seepage along the slope, and the embankment slope in slope backlog, lower intensity, seriously affect the stability of high fill substructure. Slope is not timely protection, easy to cause erosion and other diseases, and then affect the overall stability of the substructure. To this, it gives the requirements of provisions.

3.6.7 The original *Specifications* (JTG D30—2004) analysis on the stability of high embankment, do not make clear analysis of conditions, only through the provision of substructure strength parameter test using water saturated samples, the parameters used in the analysis of shallow slope stability, to consider the effect of rainfall on the stability of roadbed. In accordance with the current *Highway Codes for Earthquake Resistance* (JTG B02), the code revision increases the seismic conditions, and gives the three conditions listed in the provisions.

Rainfall affecting the stability of the substructure is mainly rainstorm or continuous rainfall. For the substructure during operation, the depth of rainfall is usually limited. China Merchants Chongqing communications research and Design Institute Co. Ltd. was obtained by different rainfall intensity field tests in Chongqing and Gansu and other places: in the absence of any substructure slope protection measures, Gansu loess substructure filling, the rainfall infiltration depth of about LRN, Chongqing shale embankment, the rainfall infiltration depth is 2–3 m. When the pavement is completed, and the substructure drainage facilities are complete, and the slope of the slope is protected by vegetation, the influence of rainfall on the substructure will be weakened. At the same time, through calculation and analysis, with the increase of rainfall penetration depth, the stability of substructure is continuously reduced. When the penetration depth is less than 3.5 m, the stability coefficient of substructure decreases slightly; when the penetration depth reaches 5 m, it reduces by about 36%; when the substructure is fully saturated, the stability coefficient can be reduced by more than half. As with all substructure affected by rain to consider the rainfall condition, not only does not conform to the reality, and will lead to the rainfall condition to control the situation, so the design should be based on the rainfall roadbed, in finite depth range to consider rainfall conditions.

3.6.8 The stability analysis of embankment involves such parameters as substructure soil, substructure fill and control level. In the long process of substructure operation, soil moisture changes, gradually tends to balance the humidity condition of its environment to adapt, is supposed to be the strength parameters of this kind of humidity conditions, but the actual situation is very complicated. The original *Specifications* (JTG D30—2004) to analyze the mechanical behavior of substructure soil, combined with the relevant test results, recommendations for normal working conditions according to compaction curve requires density corresponding to the high water content of prepared samples, test to determine the strength parameters of soil. Normally, this moisture content is 1 percentage points higher than the optimum moisture content. The *Specifications* was revised in accordance with this requirement, and for the connection with the various work conditions, the provisions of Table 3.10 of the sample preparation requirements.

The stability of substructure on slope foundation is mainly affected by soil strength parameters at the controlled level. The control level of the soil is often more complex, may be the bottom of embankment fill, foundation soil cover or underlying soft layer, may also be the contact surface of embankment and foundation or foundation soil contact surface and rock.

3.6.9, 3.6.10 There are many factors affecting the stability of high fill substructure, and they are very complicated and can not be taken into account in the stability calculation. On the basis of calculation and analysis, combined with site conditions and engineering geological analogy method, a comprehensive judgment is carried out to evaluate the stability of high fill substructure.

In the drainage facilities play a normal function, the effects of rainfall on the slope of substructure in the main part, but on the steep slope embankment, or substructure soft layer, slope and the soft layer is a key factor affecting the stability of the substructure.

The original *Specifications* (JTG D30—2004) takes into account the different ground conditions, and adopts simplified Bishop method, which considers the mean consolidation degree of soil foundation and different soil strength parameters. In order to avoid the difficulties and confusion in understanding, the standard revision adopted the usual simplified Bishop method and required the soft ground to be handled in accordance with the relevant provisions of Sect. 7.7.

The calculation formula of [Eqs. (3.6.10.1) and (3.6.10.2)] is the unbalanced thrust method implicit solution of the safety factor definition using the anti-shear strength index reduction of the traditional safety coefficient implicit in the shear strength index and transfer coefficient, through iteration. Due to the inter slice force parallel to a bottom surface of the sliding block that makes the calculation a good effect. To this end, the provisions recommend that these measures be taken when necessary.

High embankment after a rainy season, some diseases are exposed, for related treatment, conducive to settlement stability, therefore, should be high embankment through reasonable construction schedule, the settlement period set aside a rainy season.

The terrain, ground, packing the roadbed in mountainous area is very complex, although the control of domestic related research for high embankment settlement after construction, but the results have certain limitations, the settlement control standard specification also difficult to give. China Merchants Chongqing communications research and Design Institute Limited company into the highway across the high embankment of Sichuan, Guangxi and Nanning-Guilin some highway high fill substructure uneven settlement of experimental results were analyzed, obtained: to control the uneven deformation control standard, 40 mm is a proper settlement of substructure after filling the high mountain square.

3.6.14 Due to geological exploration has certain limitation, the design can not completely simulate the construction condition, the dynamic control, can according to the feedback information construction, validate and perfect design, effectively control the rate of construction, to ensure the stability of substructure, and according to the test data, analysis and evaluation of embankment settlement, reasonable pavement paving time, ensure the pavement quality and service level. Therefore, the construction monitoring and dynamic design of high embankment and steep slope embankment are very important.

3.7 High slope in cutting

3.7.2 Geological exploration of slope engineering should be carried out after geological investigation and full understanding of the geological structure of the slope. It is often difficult to achieve the desired effect by single drilling in the investigation.

The comprehensive survey of various means (inclined hole, well trough, exploratory trench, geophysical prospecting) is very effective for the identification of rock mass structure and structure. The slope survey line is usually arranged along the direction of vertical slope. The key point is to find out the lateral geological distribution of the substructure slope, and the survey area should include the areas that may affect the slope stability.

3.7.3 The mechanical parameters of rock mass slope is not easy to get, because the rock (especially the structural plane shear test site) is difficult, long test time, high cost, determination of rock properties by testing (including the structure of the shear strength index), the current project not all can do. The *Specifications* according to *Standard for Engineering Classification of Rock Masses* (GB 50,218—94) Table C.1 and combined with some test data, research results and engineering experience presents in Tables 3.12 and 3.13 for engineering design personnel. The mechanical parameters of rock slope with serious failure consequence shall be obtained by field test.

The rock standard value is to correct the test value and get the value of rock characteristic. Because there are more or less structural planes in the rock mass, its strength is lower than the strength of rock. At present, many reconnaissance units use the experience of water conservancy and hydropower systems to multiply the cohesion of rock c by 0.2, and the internal friction angle φ multiply 0.8 as the c value and φ value of rock mass. According to *Technical Code for Building Slope Engineering* (GB 50,330—2002) Table 4.5.4 and the highway department domestic experience, it think that rock c value may be used in water conservancy and hydropower system experience, the φ value of the rock mass selects in Table 3.14.

The mechanical parameters test of soil is easy to obtain, and the result is better. The in-situ shear test is recommended to obtain the mechanical parameters of the slope soil. The influence of water on slope stability mainly includes two aspects: reducing the soil strength parameters of slope and producing the water pressure of adverse slope stability. Based on these two influences, there are different methods for dealing with water in the analysis of soil slope stability: soil and water conservation, soil and water calculation. In this section, the mechanical strength index values of different methods are defined.

3.7.4 Stability evaluation of slope excavation includes qualitative judgment, the stability of the slope stability calculation, stability analysis and slope stability evaluation, development trend.

Determine the slope stability is the premise and key of slope design, it shall be carried out based on the engineering geological conditions of the slope environment full understanding and analysis on the rock slope grade and slope classification involved in this process. In recent years, the industry of our country, especially the transportation industry in the western traffic science and technology project, the rock slope grade and slope classification carried out a lot of research and achieved a lot of results, such as SMR method, RMR method, but the engineering practice test these results remains to be subjected to a wider range. Based on this, the amendment does not accept its people.

The evaluation of slope stability should follow the principle of qualitative analysis and quantitative calculation as an important assistant method. According to the engineering geological conditions, possible failure modes and the signs of deformation and failure, it is important to judge and calculate the slope stability.

3.7.5 There are many quantitative calculation methods for slope stability. The failure shape of slope is an important factor to be taken into consideration in the calculation method. There are many calculation methods for the failure modes of the same type of slopes, and the principle of the selected calculation method is given in the *SpecifiCations*.

For the calculation of the stability of the circular sliding surface slope, the Swedish strip method is adopted mostly in the highway departments in China, but the calculation accuracy is not high. The simplified Bishop method is recognized as a method with sufficient precision to meet the needs of the project. It is a powerful method recommended by the water conservancy departments and construction departments in our country. Therefore, this method is recommended in this standard.

For the stability calculation of the sliding surface slope, the Sarma and Spencer methods recognized at home and abroad are of higher accuracy, but the two methods are more complex and the designers are more difficult to use. The unbalanced thrust method is widely used in highway departments, railway departments and construction departments in our country. In view of this, the unbalanced thrust method is recommended in the *SpecifiCations*.

The numerical analysis method is a better method of slope stability, slope stability analysis of complex problems may be solved with the limit equilibrium method is difficult to solve, but so far, this method is still difficult to give general engineering and technical personnel to accept and master the slope stability calculation results and criterion. The numerical analysis of complex slope stability needs special study. Since the calculation parameters required by numerical analysis method are difficult to obtain accurately, most of the calculation results are used for qualitative analysis and evaluation.

3.7.6 The quantitative calculation results of slope stability are closely related to the factors considered in calculation, additional load and special load, which is closely related to the corresponding calculation conditions. According to the actual situation of highway slope in our country at present, the regulation of the stability calculation of cutting high slope is given.

According to the Normal working condition calculation, the parameters of the natural parameters of rock and soil slope; according to the Abnormal working conditions I calculation, the parameters of the saturated parameters of rock and soil slope; according to the Abnormal conditions II calculation, the parameters of the saturated parameters of rock and soil slope, at the same time to consider the special seismic load.

Effects of freezing thawing on the slope of the seasonal frozen region, mainly frozen stagnant water effect, the formation of shallow layer on the slope stability leads to the decrease of the slope stability in large range, the slope produces hot melt landslide. At present, the research results are not yet ripe, and this amendment does not include its provisions. Therefore, analysis of slope stability, mainly considering the influence of freeze–thaw cycles on the frozen soil physical and mechanical properties of deep rock fan, namely by freeze thawing strength parameters of saturated state, and included in the shallow frozen ice melt generated by static pressure.

3.7.7 The original *Specifications* (JTG D30—2004) analysis of motorway excavation of high slope stability control in scientific research and engineering experience, based on the relevant provisions of the slope safety coefficient of industry of our country, put forward the slope safety coefficient control standard, after ten years of mountainous motorway construction proved right. This standard revision still maintains the original standard, and has not been revised.

In the seasonal frozen area, there is no mature research result on the safety factor of the slope under freezing and thawing conditions.

Because of the different calculation methods used, the slope stability coefficient has some difference. The test results show that a large number of examples, in general, the calculation results than the unbalanced thrust method calculation results of 5%-10% Bishop simplified method; the calculation results of numerical analysis method and simplified Bishop method is close to the difference between usually within 5%; the sliding surface analytical calculation results than the unbalanced thrust method calculation results 8%-16%. According to this standard, according to the calculation of slope stability safety factor, the slope stability shall be calculated in accordance with the calculation method.

3.7.8 The excavation of high slope form and slope ratio to determine the evaluation results according to slope stability. The use of slope type in slope is beneficial to slope stability, and it should be combined with topographic and geological conditions in accordance with local conditions. It is difficult to construct steps on hard rock slopes, and the steps of consequent rock slopes in rain rich areas are sometimes disadvantageous to slope stability. At this time, it is suitable to adopt rock broken slopes without steps.

3.7.10 Water is one of the important factors influencing the stability of high slope excavation. The design should pay attention to the design of underground drainage, strengthen the comprehensive design of surface and underground drainage system, especially according to the form, occurrence of groundwater discharge modes and after excavation, groundwater state changes, according to local conditions, choose the suitable scheme for drainage.

3.7.11 Dynamic design is one of the design content of high slope excavation. The designer should timely track and grasp the real geological characteristics, excavation slope deformation, reflected in the measured stress data, check and supplement to the original design, perfect design, to ensure the safety and rationality of design engineering.

Geological data is the basis of the design, but the geological condition is complicated, affected by various factors, geological exploration data and ensure the accuracy rate is low, and the main conclusion errors caused by the slope failure phenomenon is the case. Therefore, it is necessary to make supplementary investigations on the high slopes with complicated geological conditions, collect geological data, check and verify geological survey conclusions, and avoid the engineering accidents caused by the failure of the survey conclusions.

Field monitoring is a highly technical work, it plays an important role in the correct implementation of engineering design, and is also an important basis for ensuring the safety of construction or emergency management. Therefore, in the design document, the entire monitoring system, procedures, content, technical requirements should be clearly defined.

3.8 Rock embankment

3.8.1 Rock filling has good engineering properties such as good compactness, high permeability, large filling density, high shear strength, small settlement deformation and high bearing capacity. It has been widely used in highway engineering construction. But the stone lithology on rock fill embankment structure affect the performance of large, expansive rock, soluble rock and salt rock, in the water environment and with age growth, the engineering properties will deteriorate, the poor stability, easy to produce roadbed disease, is not used for embankment filling.

The quality of the stone filled embankment is closely related to the performance of the construction equipment. When the power of the compaction machine is small, the filling stone can not be further broken, and the compaction effect is poor. When the power of bulldozer is small, it is difficult to shift and pave the large size rockfill, and the surface is further broken and compacted. The test shows that the static weight below 12 t vibration roller, rolling in the hard strength more than a stone, is difficult to make a great impact on the smooth surface, elegant effect. Therefore, the construction method of large power bulldozer and heavy compaction machine is recommended in the *SpecifiCations*.

The construction of rock filling embankment with impact compaction and dynamic compaction may guarantee its construction quality and mature engineering experience. When conditions permit, the design can adopt the construction scheme.

3.8.2 The original *Specifications* (JTG D30—2004) highway embankment construction experience summary and research achievements, from the *Water Conservancy and Hydropower Engineering Geology Investigation Code* (GB 50,287) according to the classification standard of the uniaxial compressive strength, according to the requirements of the engineering properties and construction technology of filling materials for the angle of the given rock fill embankment classification of highway (Table 3.16); hard rock (more than 60 MPa), hard rock, soft rock (30–60 MPa) (5–30 MPa). For extremely soft rock with strength less than 5 MPa, the construction and compaction characteristics can be taken into account according to the soil packing. Practice has proved that the above classification is appropriate. The *Specifications* revision maintains this classification standard.

3.8.3 The original *Specifications* (JTG D30—2004) summary of Fujian Fuquan Beijing Zhuhai Motorway, Guangdong Motorway, Guangxi Motorway Liuzhou Guilin granite, limestone, red sandstone embankment test road engineering experience and achievements in scientific research, put forward different rock types of paving layer thickness, the maximum particle size and porosity of the compaction quality standard.

The porosity as the detection index of compaction quality is unique, but the need for pit (maximum size of 1.5–2 times) bag test, testing is difficult. In recent years, experimental research and engineering practice have been carried out on the differential settlement of compacted soil as a testing index. The compaction settlement of rock filling is closely related to the actual power dense state and roller is reflected by the difference, roller hard rock with small power, the differential settlement is not representative of the actual state of rock filling compaction can meet the engineering requirements. The key to quality control is that the power of the roller matches the strength of the filling stone, that is, only the heavy vibratory roller can be used to ensure the quality of the project.

The revised standard, in order to ensure embankment compaction quality, and easy construction compaction quality detection, provide the standard of compaction quality of embankment with porosity as a control index, the porosity and compaction quality control combined with differential settlement or construction parameters. In actual engineering construction, the test road determines the compaction settlement difference and the control standard, and the porosity index is also checked to verify it.

The difference between the compaction and settlement is the difference of the elevation of each measuring point after rolling by the heavy vibratory roller (above 14 t) and the rolling parameter (strong vibration, speed below 4 km/h). It is suggested that the following standards should be adopted for compaction settlement differential test: the average difference between compaction and settlement shall be no more than 5 mm, and the standard deviation shall not be greater than 3 mm.

3.8.5 The slope of rock fill embankment is often the weak link of spreading and compaction, and it is difficult to make the slope close and smooth by conventional method. Therefore, the rock slope of hard rock embankment needs to be cemented by side slope.

3.8.6 Weathered rock and soft rock embankment in highway operation period, in the climate (rainfall, evaporation, dry wet cycle), surface water and groundwater under the influence of the physical and mechanical properties of attenuation, especially after immersion of soft rock will soften, the shear strength and bearing capacity, caused by soft rock embankment the uneven settlement of deformation and cracking, even instability. The key of preventing and controlling soft rock substructure diseases is to control the moisture stability of soft rock substructure and avoid the deterioration of soft rock properties under the influence of climate, environment and water. According to the achievements in scientific research and relevant engineering experience, measures such as edging, sealing or reinforcing the bottom of the substructure and setting up the drainage cushion at the bottom are put forward in this code.

3.9 Light weight material embankment

3.9.1 A lightweight embankment is an embankment filled with a material that is seriously smaller than a fine-grained soil. There are many types of lightweight materials used as highway substructure filling materials. From the point of view of technical maturity, they are mainly made of geotextile foam, foamed light soil and fly ash. The purpose and function of them are basically the same. The *Specifications* was revised to incorporate it as a 'lightweight material embankment'.

The soil particles in the process of forming foam expansion formed many uniform closed cavities, this structure determines its light weight, pressure, water and many other excellent engineering properties, for the elimination of the substructure construction of embankment and embankment in general on the soft foundation settlement or differential settlement has significant effect. At the same time, the foamed plastic block has the advantages of independence and small lateral deformation, which can greatly reduce or eliminate the lateral pressure on the abutment or retaining wall structure. In addition, the thermal conductivity of the geotextile foam is lower, and it has a remarkable thermal insulation performance. It may be used to reduce the freezing depth of substructure and to reduce the melting of permafrost soil.

Foam light soil, also known as the bubble mixed light soil, is a new type of light filling materials of highway construction field, has the advantages of light weight, severe and adjustable intensity, gravity, upright and easy excavation and construction characteristics of convenience etc. The filling engineering in order to reduce the load or earth pressure for the purpose, has been widely used in soft foundation embankment and soft foundation abutment, road widening embankment, steep embankment, underground structure top load reducing backfill, soft soil foundation and foundation treatment, emergency repair and rapid collapse of embankment engineering in cold regions. Foam light soil first by the Japan road team in 1986 applied in road engineering, Guangdong Province imported from Japan in 2001, and has been widely applied in domestic highway construction, has become a mature embankment construction material. Foam lightweight soil is a kind of lightweight concrete solidified by adding foam in cement base slurry. In practical engineering, its raw material is mainly composed of cement, water and foam, except for special performance requirements. The admixture can be selected according to performance requirements and economy. For example, in the fly ash is rich and cheap, may be mixed with fly ash; in need of higher strength, may be mixed with fine sand and other admixture; in aeolian sand rich areas may be mixed aeolian sand; water requirements in the underground water level when filling may be mixed with waterproof agent material.

At present, the use of fly ash embankment of fly ash is mainly wet ash (Chi Hui), the wet ash, low calcium fly ash as silicon aluminum, high calcium fly ash and bottom ash dry ash and sulfur calcium were lack of practical experience and application, therefore, the *Specifications* is only low calcium fly ash on silicon aluminum.

3.9.3 2 The most important indexes of foamed lightweight soil embankment are wet strength and compressive strength of construction. Japan Road Association FCB, provisions of the design and construction method of pointer foam light soil properties are shown in Table 3.2, the Ministry of Housing and Urban Industry standard *Bubble Mixed Light Soil Filling Engineering Technology*(CJJ/T 177—2012) the provisions of the foam light soil properties are shown in Tables 3.3 and 3.4. In recent years, foamed lightweight soil has been widely used in highway engineering. The foamed lightweight soil is mainly composed of cement, water and foam. The wet and compressive strength of the filling site has drawn lessons from the standards of the Ministry of Housing. The whole strength and stability of the embankment are better, but the shrinkage cracks are more. Guangdong province to broaden the Beijing

Table 3.2 Performance index of foamed lightweight soil of Japan road public corporation

Filling site	CBR (%)	Single axis compressive strength (kPa)
The upper part of the roadbed	10	1,000
The lower part of the roadbed	5	500
Highway part	2.5	300

Table 3.3 Performance index for substructure filling

Depth below the pavement (m)	Minimum strength level		Minimum severity class
	Urban motorway, highway, first class highway and trunk road	Other class of highway	
0–0.8	CF0.8	CF0.6	W5
0.8–1.5	CF0.5	CF0.4	W3
>1.5	CF0.4		

Table 3.4 The performance indexes used to calculate of filling below the water level

Below calculate water level (m)	Minimum severity class	Minimum strength level
≤ 3	W6	CF0.8
> 3	W8	CF1.0

Zhuhai Highway two North Ring Road of Guangzhou Taihe exchange base in the project, the foam lightweight soil mixed with sand mix and the engineering properties and construction technology were studied, using cement:sand = 1:2, moist unit weight 10.8 kN/m^3, current value 180 mm, the compressive strength reached 2.85 MPa, achieved good effect. According to the engineering experience and research, puts forward the requirements of foam light soil performance index Table 3.21 for substructure of this code; in order to improve the highway embankment foam durability, reduce shrinkage cracks, improve the foam light soil performance index of highway traffic, especially heavy, and require the use of sand mix.

The engineering requirements should be clear when the indexes of frost resistance of foamed lightweight soil are determined, and the variation of the intensity and intensity of the lightweight soil is determined by experiment, and then the relevant indexes are designed. If no test data, it designs according to the requirements like heavy loss rate is not greater than 5%, the compressive strength loss rate is not greater than 10%.

3 The specification of the fly ash is a silicon aluminum low calcium fly ash, equivalent to the American Standard (ASTM C618-87) F in fly ash. The standard is formulated for technical requirements for fly ash used as common cement concrete additions. The maximum amount of ignition loss is 6%. If there is any test data available, F fly ash with the maximum amount of heat loss of 12% may be allowed, but the embankment filler is not specified. The current *Technical Code for Construction*

of Highway Pavement Base (JTJ 034—2000) stipulates that the loss of fly ash shall not exceed 20%. As an embankment filler, it is feasible to use the same provisions as the base material.

3.9.4 When lightweight materials such as foamed plastics and foamed lightweight soil are exposed directly, the light material can cause aging and other damages under the influence of environmental factors. Geotechnical plastic foam in the ultraviolet irradiation aging rodents, easy to produce harmful substances, such as fire will cause damage to the soil block foam; foam light soil exposed prone to carbonation metamorphism, and lead to a significant reduction in strength and weathering; fly ash exposed, in the evaporation, water loss dry fly ash without cohesion, the slope stability is poor, the rainfall will make the ash slope more serious erosion. Therefore, this standard stipulates that the design of lightweight materials embankment shall adopt effective protective measures, and lightweight materials shall not be exposed directly.

In the design of lightweight embankment, slope embankments and vertical embankments are often used to prevent light materials from being exposed directly. Slope embankments are slope protected by soil, while vertical embankments are made of concrete face panels or retaining walls. Foam light soil embankment with vertical setting, concrete revetment wall protection layer, the precast concrete wall panel height is less than 3 m; the height of more than 3 m, usually using cast-in-place reinforced concrete retaining wall.

Lightweight materials mainly is in order to reduce embankment weight, reducing the ground stress and settlement as the main purpose, the filling thickness according to the calculated settlement calculation, should follow the 'foundation stress equivalent principle' or 'settlement control' principle.

3.9.5 Soil embankment foam protecting layer reinforced concrete slab to increase soil block foam integrity, better dispersion of vehicle load and overburden pressure, avoid damage of construction loads on soil block foam, and prevent harmful substances soaked soil block foam. Reinforced concrete slab protection layer generally adopts diameter of phi $\phi 6$ mm, 15 cm × 15 cm steel mesh.

3.9.6 Under the influence of the environment, the foamed lightweight soil will produce shrinkage cracks due to the change of humidity and temperature. In order to prevent the water from seeping into the foamed soil, reduce the shrinkage cracks and prevent the cracks from reflecting on the road, it is necessary to set up metal mesh and polyethylene geomembrane on top of the road.

3.9.7 The saturation strength of fly ash is very low, and the strength and stability of the fly ash embankment will be insufficient when the surface water and groundwater penetrate into the interior of the fly ash embankment. Therefore, the design of fly ash embankment should consider the effect of capillary water separation, and set up drainage cushion and isolation layer at the bottom of embankment, so as to prevent long-term water immersion embankment foundation. Fly ash embankment slope on both sides need to set the drainage ditch edge, and the nonwoven geotextile as filter layer to prevent the blockage of drainage ditch.

Table 3.5 Control limits of hazardous substances for landfill waste

Project	Control limit (mg/L)	Project	Control limit (mg/L)
Organic mercury	0.001	Zinc and its compounds (in total zinc)	75
Mercury and its compounds (in total mercury)	0.25	Beryllium and its compounds (in total beryllium)	0.20
Lead (in total lead)	5	Barium and its compounds (in total barium)	150
Cadmium (in total cadmium)	0.50	Nickel and its compounds (in total nickel)	15
Total chromium	12	Arsenic and its compounds (in total arsenic)	2.5
Hexavalent chromium	2.50	Inorganic fluoride (excl. calcium fluoride)	100
Copper and its compounds (in total copper)	75	Cyanide (as measured by CN)	5

3.10 Embankment with industrial waste material

3.10.1 It is of great significance to save the land and protect the environment by using industrial waste residue to fill the embankment. However, some abandoned slag due to some enrichment of trace elements in the forming process, has a certain toxic, corrosive or radioactive, have potential adverse effects on the environment and human health, the design should pay full attention to the contents of these elements, especially the content of trace elements in industrial waste leachate. In the *Technical Requirements for the Construction of Hazardous Waste Safety Landfill Disposal Project* issued by the State Environmental Protection Administration in 2004, the waste control limits of the direct human field (non-pretreatment) landfill are shown in Table 3.5. When industrial waste residue contains harmful substances, it cannot be used as embankment filler.

Some industrial waste (coal gangue and cinder) containing a certain amount of sulfate ions or leaching solution is acidic, aggressive, some of the structures so permanent for distance concrete structures and metal component assembly structure and pavement structure layer within 0.5 m embankment in the soluble salt content and pH index, to meet the concrete structure, metal component and pavement durability design requirements. For example, the UK highway standards, within the concrete, cement materials and cement stabilized soil material 500 mm, water solution of substructure not sulfate content exceeds 1,500 mg/L; within the permanent structure range of 500 mm assembled on the metal component, water soluble sulfate content of roadbed filling cannot exceed 300 mg/L.

3.10.2 Blast furnace slag and steel slag have the advantages of high carrying capacity, good solidity and high strength, but they also have the characteristics of powdering and expansion. The main factors affecting the stability and stability of blast furnace slag and steel slag are the content of free calcium oxide (f-CaO) and MgO, as well as the particle size and storage period. The natural aging digestion method which uses outdoor storage for a certain period of time to make it expand and break is an effective measure to solve its stability. The test shows that the new slag with a storage period of less than one year has a larger expansion rate; the storage period is more than one year, and the slag expansion rate is smaller. According to the research results and engineering experience, the specification of the storage of more than one year Chen slag, and puts forward the swelling rate, autoclave pulverization rate, the content of free calcium oxide control standard.

The water absorption, disintegration, expansion and spontaneous combustion of coal gangue have great influence on the stability of substructure. The coal gangue can be classified according to the plastic index and value. The results show that the coal gangue with plasticity index greater than 10 usually contains more unstable components such as montmorillonite and illite, and the swelling rate is relatively large, so it cannot be used as embankment filling directly. The burned coal gangue compared with coal gangue, tend to have better stability (such as small expansion rate); but has not fully oxidized coal gangue in the coal and air chemical reaction, generating a large amount of gas and heat, resulting in volume expansion, caused by the deformation of substructure and pavement cracking and other diseases. Therefore, this standard stipulates that the gangue which is not fully oxidized and aged and whose plastic index is greater than 10 should not be used directly for filling highway and first class highway embankment.

Coal gangue with poor properties refers to the coal gangue with high carbon content, more than 20% loss of ignition, a plasticity index greater than 10, a free swelling rate greater than 40%, and a CBR strength that does not meet the requirements of this code.

3.10.4 In view of the industrial solid waste leaching material can cause adverse effects on the environment, and the coarse slag disintegration, fine slag particles migration and piping will cause adverse effects on the substructure stability and long-term performance, the specification of industrial solid waste leaching water shall not be used in embankment section and flood inundation area.

3.10.5 To prevent the adverse effects of industrial waste residue on the environment, closed embankment structures are usually used. Industrial waste residue embankment consists of main part of the embankment (industrial waste), slope protection and capping layer (cohesive soil or other materials), as well as isolation layer, drainage system and other parts.

4. Substructure Drainage

4.1 General requirements

4.1.1 Water is the major cause of substructure defects. Substructure defects can be resulted from imperfect and irrational waterproof and drainage system, shortage of cross section and premature failure caused by improper selection of drainage facilities. Designers must pay attention to substructure drainage design.

The substructure drainage consists of surface drainage and underground drainage. Surface drainage is to drain the surface runoff, slope rain and other surface water close to the substructure that will impact the substructure stability. Underground drainage is to drain the underground water flowing to the substructure or lower the underground water level.

Rational arrangement of drainage facilities and the design of cross section are very important in the drainage system design. The design shall be carried out according to road classification, meteorology, hydrology, terrain, geology and set of bridge, culvert and tunnel to arrange drainage system that kinds of drainage facilities working together to drain water out of substructure and maintain a dry or medium wet state.

China covers a large area that each part has special climatic and geologic conditions. Substructure drainage usually influence each other with surface drainage, substructure protection, foundation treatment and should be considered together.

4.1.2 Surface water of road boundary usually are not allowed to flow into the bridge, tunnel and their drainage systems. This is to avoid surface water carrying soil and trash which can plug the drainage system and scouring the abutment and cone slopes to the bridges and tunnels to impact the safety.

When the tunnel exit direction of cut substructure is uphill, reverse slope drainage usually can be set along the route. When the cutting and longitudinal slope out of tunnel are too long, reverse slope drainage will has large amount of engineering and its outlet position and elevation are difficulty to determine. It is obviously irrationally to mandate the drainage being reverse slope. When the length of tunnel is less than 300 m, drainage is difficult and the mud content and amount of surface water out of tunnel is little, the water in cutting edge can be drained through tunnel. The cross section should be enough when the water be drained through tunnel. Grit chamber and geogrid also should be set to avoid carrying soil and trash into drainage facilities in the tunnel.

It is needed to set intercepting drain, chute, stilling basin and other surface drainage facilities with slope protection measures on both sides of the substructure to avoid scouring on bridge abutment substructures and cone slopes.

4.1.3 For sections of low-fill, shallow excavation, or drain difficulties, the road surface and subgrade work area are usually close to the underground water table. The long-term performance of substructures in these sections will change because of the joint action vehicle loads and water. This can easily lead to subgrade and pavement damage. Drainage design is a major problem in sections of low-fill, shallow excavation, or drain difficulties.

4.1.6, 4.1.7 Road construction will definitely impact the old drainage system along the route. Extreme rainstorm weather is also important to be considered in drainage design and construction because it has features of suddenness, high strength and serious damage. Temporary drainage facilities are necessary for ensuring construction quality of substructures, pavements and bridges and protecting environment along the route. Temporary drainage facilities for the construction site should be combined with the permanent drainage facilities to save investment and convenient construction.

4.2 Surface drainage

4.2.1 Hydrological computation of substructure may use methods in *Specification for Drainage Design in Highways* (JTG/T D33) or *Highway Design Manual Subgrade.* According to foreign data, rainfall return period of substructure design in this specification is 15 years for motorway and class-1 highway and is 10 years for classified highway. The choice of hydrological computation formula should be according to the position, effect and catchment area.

The dimension side ditch, intercepting ditch, drainage ditch, drop water and chute should be ensured to fit the total design flow. Flow velocity in ditches should be in the section between maximum and minimum allowed flow velocity.

4.2.4, 4.2.6 According to explanation in *Compilation Regulation for Highway Engineering Standards* (JTJ 002) and *Specifications for Drainage Design of Highway* (JTG/T D33), the *Specifications* defined side ditch and drainage ditch more clearly.

Side ditch includes cut side ditch and embankment side ditch. Side ditch is usually located outside soil shoulder and berm and used to collect and drain water out of pavement, shoulder and slope. Widely used cross sections of side ditch include triangle, shallow dish shape, U shape, trapezoid, rectangle, rectangle with cover plate, blind ditch and so on. The shape choice is needed to consider terrain condition, slope height, water catchment area, drainage functions and the impact in environment and safety. Triangular and shallow dish shaped side ditches are suggested when slope height and water catchment area are not big. The size of cross section should be determined through Hydrological and hydraulic calculation according to terrain, water catchment area, rainstorm intensity, filling and excavation condition and local experience.

Different protection and reinforcement measures should be chosen in different situations in scour protection design of slope. This specification defined maximum allowed flow velocity of different types of open trench as the basis of reinforcement design of open trench. Environmental compatibility and durability need to be considered when choose side ditch reinforcement measures. If the flow velocity in side ditch is lower than maximum allowable velocity, it is prefer to use plant revetment. If the flow velocity is bigger than maximum allowable velocity, it is suggested to choose indirect reinforcement measures include gravel, slice rubble and pebble gravel, or direct measures include dry-laid rubble, mortar-laid rubble and cast-in-place concrete.

4.2.5 The usage of intercepting ditch is to intercept surface water from the slope above cutting and prevent seepage of the water in it to impact the slope stability. Therefore, the distance from the intercepting ditches on the upside from embankment to the toe of the embankment slope should be considered in the two aspects. If the distance is too big, much water could not be intercepted and will scour the slope. Both soil condition of cutting slope and harm to highway operation after slope collapse should be considered in design. The distance from the intercepting ditches on the upside from embankment to the toe of the embankment slope shall not be less than 2 m.

Intercepting ditch can be made into embankment ditch, cutting topside ditch and platform ditch according to the filling and excavation condition and position. Position and channel number of intercepting ditch are very important to be determined by flow calculation according to hydrological, topographic and geological terrain. Intercepting ditch need anti-seepage facilities to avoid slope failure. The outlet shall extend beyond the substructure and set with energy dissipation facilities to ensure the stability of slope and substructure.

Type of cross intersection and impact in environment shall be considered in the design. Intercepting ditch usually do not set when the catchment area is little on cutting top.

4.2.7 Drop water and torrent chute are mainly used in steep slope areas to dissipate water energy and slow the water velocity. They are generally used in mountain highways as drainage facilities. The type of cross section are usually rectangle and trapezoid. Intake and outlet are the key points of water erosion failure that well set protection and reinforcement are needed.

4.2.8 Evaporation pond is suitable only in areas like northern China that are dry, easy to evaporate and difficult to drain. Volume of each evaporation ponds shall be determined by its longitude spacing through hydrology, hydraulic and evaporation calculation. Appropriate protection and strengthening measures should be taken to prevent pedestrians from accidents.

4.2.9 *Specification for Drainage Design of Highway* (JTG/D 33—2012) defined 'sensitive water quality section' and its drainage design requirement.

The sewage from pavement is mainly suspended solid and petroleum. Precipitation method is suggested to dispose the sewage because its oil content is lower than other sewage. Oil–water separation pond is seldom used in China. The design may be according to *Code for Design of Outdoor Wastewater Engineering* (GB 50,014) and *Integrated Wastewater Discharge Standard* (GB 8978).

4.2.10 Down channel is one of the key measures to ensure the function and safe operation of undercrossing. The main problem of down channel design is that there is still no standard about the surface runoff and drainage system. *Code for Design of Outdoor Wastewater Engineering* (GB 50,014—2006) stipulates the drainage design recurrence period of down channel should be more than 3 years. This specification stipulate the drainage design recurrence period of down channel should be more than 5 years in the consideration of highway features, current climate and environment change especially frequency and intensity of rainstorms.

According to the experience and researches, the *Specifications* gives the operating condition and scope of mainly used drainage methods in down channel (gravity drainage, pumping station drainage, seepage well drainage, evaporation pond). The choice of these methods should be based on the specific conditions. Gravity drainage is preferred to use because of the capacity and maintenance of pumping station drainage, seepage well drainage and evaporation pond.

4.3 Underground drainage

4.3.1 Hydrogeological parameters are important basic materials in underground drainage design. Before the design of underground drainage, based on the existing relevant data, the field engineering geological and hydrogeological surveys, explorations and tests need to be carried out to find out hydrogeological conditions and obtain related hydrogeological parameters.

4.3.2 The types of underground drainage facilities are abundant. The using condition and scope, drainage capacity and long-term effectiveness of these facilities are different too. The *Specifications* gives the choice standard according to the condition of underground water.

4.3.5 Underdrain may be divided to rock-fill underdrain, non-fines concrete underdrain, pipe underdrain and tunnel underdrain according to their materials and structures.

Rock-fill underdrain is usually used in sections with little underground flow and short length of underdrain. It is easily to be filled. Pipe and tunnel underdrain can be used in sections where underground flow is high and the drainage distance is long. Tunnel underdrain shall be preferred to use. Pipe underdrain is difficult to construct and ensure the quality. Tunnel underdrain is usually used to replace rock-fill underdrain and non-fines concrete underdrain.

Slope underdrain and revetment underdrain are mainly used to dry the wet soil cutting slope and drain perched water and spring water on parts of the slope. The slope is covered with dry-laid rubble to keep the slope dry and stable.

Filtrationgeotextile and anti-seepage geotextile used in underdrain is usually chosen according to hydrogeological conditions and used position through *Geosynthetics* (GB/T 17,638–17,642).

Non-fines concrete underdrain may be used as filter layer or seepage ditch. It is a new drainage facility used in underdrain recently years. Non-fines concrete underdrain is used as permeable well wall or ditch wall to replace complex filter layer and seepage hole equipment. Non-fines concrete underdrain may bear limited loads and has good permeability and filtration. It is also easy for construction and can save material. It is effective to use precast no-fines concrete slab in gravel and coarse middle sand aquifer as filter layer. In fine sand aquifer, geotextile is suggested to set outside of precast no-fines concrete slab as filter layer to avoid fine soil grains filling pores in non-fines concrete slab.

4.3.6 Upward-inclinedly drilled drainage holes (UDDH) is an effective method to use small diameter drainage pipe to drain deep groundwater. It is usually used for the discharge of water in fixed aquifer in slope and the centralized groundwater exposure on slope. UDDH is usually arranged in groups. It has been widely used in mountain highways in China. The longest drainage hole is 50 m.

Diameter of UDDH is usually 75–150 mm. Its elevation angle is suggested more than 6°.UDDH shall extend into groundwater-enriched zone or potential sliding surfaces. Permeable pipe in UDDH has the diameter of 50–100 mm. Inlet of UDDH and seepage pipe section shall be wrapped with permeable geotextile to prevent blockage of seepage holes.

4.3.8 Drainage tunnel can be used to intercept and lead the deep underground water. Drainage tunnels usually work with percolation well and pipe group to drain groundwater in complex stratum with multi-layered aquifer. Drainage tunnel should be embedded in the stable strata near the main aquifer need to be intercepted. The top of drainage tunnel in sliding area shall be below the sliding surface (zone) or the stable stratum not less than 0.5 m.

The longitudinal slope of drainage tunnel should be determined by groundwater depth, hydraulic gradient and elevation of the outlet. Width of the cross section of drainage tunnel is usually controlled by construction needs not drainage discharge. The work space for one worker after the reinforcement of tunnel wall should be considered when artificial construction. Construction ventilation also need to be considered to widen cross section in long drainage tunnel.

5. Substructure Protection and Retaining Structures

5.1 General requirements

5.1.1 Substructure is long term influenced by natural factors. Rock and soil's physical and mechanical properties will change under the adverse hydrological action. This will result in substructure failures. Substructure protection, retaining and reinforcement are necessary for keeping the stability and strength of substructure.

5.1.2 The main functions of slope protection are to protect slopes against impacts of rain erosion, weathering, temperature difference and humidity change and prevent or slow weathering, broking and exfoliation on the surface to keep the stability of substructure slopes. Slopes must have enough stability in design without considering lateral pressure of slope stratum.

Flora protection can protect and improve the environment. It is preferred to use on slopes which are suitable for plant growth and have the slope grade less than 1:1. Because plant growth is slow in the early stage and the protection ability of flora protection is weak, mortar rubble or concrete aggregate should be set on soil slopes with large height and length. For complete stable slope consisted of weak/ micro/ unweathered hard rock, protection can be spared.

5.1.4 Rules below are proposed from mountain highway construction experience in China and relevant data.

(1) The design of retaining structures need to ensure stability, solidity, and durability of subgrade in various design load combinations.
(2) Suitable types of subgrade retaining not only gravity retaining wall may be chosen according to topographical, geological and hydrological conditions combined with engineering technology through principles of technology reliability, economic rationality and environmental harmony. All the choice need to meet requirements of safety, economy and easy construction and maintenance

(3) Materials of retaining facilities need to be chosen according to the structure type. Gravity retaining wall can use mortar rubble or concrete aggregate locally. Other retaining facilities can use materials, including geosynthetics, metal straps, anchored retaining wall and tensile reinforcement in anchor plate retaining wall, more than reinforced concrete according to the structure. These materials need to meet the requirement of durability and corrosion resistance because materials in filler are easy to be corroded by water and other chemical compositions.

5.2 Slope protection

5.2.1 Slope protection is one of the measures to protect stability of slope, improve road improvement, protect environment and prevent soil erosion. Table 5.1 gives widely used types of slope protection and the applicable conditions. It is needed to consider the applicable condition and impacts on the environment when choose suitable type of slope protection. Flora protection has the priority when climate and soil meet the requirement.

Masonry protection (shotcrete support, hanging nets, dry-laid rubble slope protection, mortared rubble slope protection, facing wall) has the main problem that it jars with the surroundings. It needs to be used as less as possible especially in Motorways and Class-1 highways and tourist highways. If masonry protection has to be used, details need to be improved to fit the surroundings.

Cover protection and thumped surface protection in old *Specifications* (JTG D30—2004) have been seldom used now. The *Specifications has* deleted relevant contents.

5.2.2 There are many types of flora protections. It is needed to choose suitable type of flora protection according to local soil, slope height and the climate. Grass seed should be perennial herb and combined with shrub to improve the erosion resistance and durability.

Spray-sowing is to cover or replace some thickness of soil or soil mixture (including soil, organic matter, fertilizer, water retention materials, insecticide and grass seed) which is suitable for plant growth on slopes where plant is difficult to grow. The thickness is determined by slope grade, rainfall, rock mass structure, lithology and the kind of plant. The thickness can influence the growth of plant. The plant may die easily in torrid seasons. The thickness for spray-sowing needs to not be less than 0.10 m according to the application situation.

5.2.3 Flora protection in mortar rubble frame is one of main measures of slope protection that can not only intercept or slow flow on slope to prevent erosion but also improve landscape.

Flora protection in mortar rubble frame is usually used in slopes with grade of 1:0.5–1:0.75 in pluvial regions in south China. Rainfall can result in topsoil slip and frame void in the early stage of plant growth and damage slopes. This problem must be considered in design. Soil fixed measures or shrub transplanting in the frame are needed when necessary.

5.2.6 Protection wall is widely used in cutting protection. It is usually combined with upward-inclinedly drilled drainage holes and underdrain to avoid deformation failure of slopes. Protection don't bear any other loads and soil pressure from the back of the wall except its gravity. Consider the landscape, the height of protection walls of Motorways and Class-1 highways should not be too high.

5.3 Substructure protection along riverside

5.3.1 Substructure along riversides often collapse or break because of flood. Scour prevention of substructure is the main measure against washout of mountain highways.

Erosion protection has direct and indirect types. Direct protection is to prevent substructures and river banks from direct damage of water. Slope and toe are the points of protection. Protection and reinforcement facilities are directly set on slopes and river banks to defend water erosion. Indirect protection is to use measures like river diversion to eliminate or slow down the direct damage of water on substructure and river banks. Water near the river bank will slow and result in sediment deposition as another kind of protection.

River diversion is to change the flow direction by groins to reduce the erosion on substructure. It is usually used where the riverbed is wide, erosion and deposition have a balance, and flow properties are easy to change.

When the substructure occupies too much riverbed area or the flow directly threat the safety of substructure, river course change projects can be adopted if it meets the requirement of topographic and geologic conditions. River course change projects are not suggested to be adopted in canyon, debris flow sections and instability reaches.

Table 5.2 gives widely used types of substructure protection along riverside and the applicable conditions. Protection measures need to be chosen carefully according to features of reaches. Measures may be used alone or together.

5.3.2 The quality of foundations of scour protection works is the key of construction. According to practical experience, failures of scour protection works usually result from foundations which have been washed out. The foundation of scour protection works should be buried no less than 1 m below the scour depth or should be embedded within the bed rocks.

Measures to avoid scour can be classified into vertical and horizontal measures according to the properties. The vertical measures are to bury the foundation below the scour depth. Pile foundation or open caisson can be considered when the scour depth is large because the open cut construction is difficult in this depth. Horizontal measures are to tile flexible facilities on the riverbed or stack bulk materials in front of the major project. When the riverbed be scoured, these projects, such as riprap, gabion, submerged dam and concrete slabs, can sink and protect the foundations. Horizontal measures can only be set in second or below grade highways or used as the auxiliary measures of vertical measures.

5.3.3 Another cause of washout damage of mountain highways is that hydrologic conditions are changed because of the worse flow condition resulted from causes that the plane shape of the axes of protection projects cannot fit the protected river banks or the joints of the both ends of the protection projects with river banks are not smooth or the buried depth of the ends is not enough. The *Specifications* presents the basic requirements of the combination design of horizontal protection projects with river banks.

5.3.4 Water diversion route is the key point of water diversion structure. Water diversion route is the new bank line made by changing the river flow away from the substructure with water diversion structures. It is the basis of arranging diversion structures. The design of water diversion route needs to fit the requirements of the predetermined river axes and bank lines. The disadvantage scour resulted from water diversion route need to be avoided.

Impacts in river flow, terrain of banks and riverbeds, upstream and downstream banks need to be considered in diversion route design. The plane shape of diversion route must fit the shape of river banks need to be protected. Flow in diversion sections need to be smooth to avoid scour in farmlands, villages, upstream and downstream substructure. The starting point of diversion route needs to be chosen according to hydrogeological conditions and be set in sections where has good geological conditions and can ensure the flow can be changed as diversion direction. The end of diversion route needs to be connected with natural downstream river bed smoothly to avoid disturb the downstream.

Water diversion projects should not be used in sections in narrow river valleys because the riverbed is narrow and flow velocity is high.

5.3.7 Groin is a kind of river improving structure which can change rivers obviously. It is usually used in instable rivers have frequently changed width, depth or flow direction. The length of the groin is determined according to river diversion route. The shorter groins, which should be mainly used for protection, can only influence the surrounding water without obvious change in flow near opposite bank. The longer groins are mainly used to jet flow or change the flow direction away from protected banks.

Groin is mainly used to prevent retaining wall and slope from scour on shallow foundations. Overflow dam and submerged dam are the mainly used groin. With the change of the intersection angle between the groin and flow direction, the impacts on flow structure and the recirculating zone size and the sedimentation on the back of groin will change too. To basin bam and upward trajectory bucket type dam, water will flow in the direction perpendicular to the direction of the groin axes after overflowing the groins. This will result in horizontal axial circumfluence along dam body. The flow direction is to the banks. Sedimentation will happen near the banks. The circumfluence area of upward trajectory bucket type dam is longer than that of downward trajectory bucket type dam. This specification suggests the intersection angle between the groin and flow direction should be less than or equal to 90°.

Root of the dam is the place where river banks combined with beaches. It is mainly impacted by swirling flow and is easy to be damaged by flow. The damage can fail the function of groins. The joint between dam root and the slope needs to be buried firmly in the old bank.

5.3.8 Longitudinal dam, which is also called diversion dam, has the function of water diversion. It almost do not change the old flow structure. Longitudinal dam is mainly used on narrow riverbed where is not allowed to be occupied much or areas where quantities are too large to construct groin or slope protection.

Grid dams are usually set in the shape of grid between longitude dams and the slopes. Grid dams usually work with longitude dams to promote sedimentation and defend scour on banks and slopes. The distance between grid dams shall be determined by the principle of slowing flows between grid dams.

5.3.9 The targets of river diversion are to change the flow direction which scour the substructure directly, wide the riverbed if the substructure occupied some riverbed, excavate the beach to protect substructure or cut and straighten the curve for route or bridge arrangement. River diversion projects need to be treated carefully. The project can only be adopted after techno economic appraisal if it is really effective.

5.4 Retaining wall

5.4.1 Retaining wall is the structure used for supporting the soil or slope to avoid deformation instability. Table 5.3 classifies different types of retaining wall according to the structure style and the applicable conditions.

Retaining walls need to be chosen according to the techno economic appraisal of topographic and geological conditions, load conditions, height, buried depth of foundation, requirements of the environment, construction difficulty, engineering cost and land use. The comparison of schemes included two aspects. One is the comparison between retaining wall and other retaining structures. Another is the comparison between different types of retaining walls.

5.4.2 Limit state of a function means that the retaining wall has been above a special state and lose the function under common maintenance.

There are two types of limit states: (1) Carrying capacity limit state. (2) Serviceability limit state.

Limit state can be known as the maximum load state related to safety. If plastic deformation happens in a retaining wall and results in significant change in geometry, it can be seen in carrying capacity limit state because safety has been in a risk. The serviceability limit state of retaining walls is related to the stability and durability. It means that functions of the retaining wall or some parts have reached some limit state. This is only related to the work condition and performance of retaining walls. Some constraint conditions, such as the crack width, deflection of the walls and so on, are needed.

This specification uses the partial factor method of limit state as the design method. Basic variables in Eqs. (H.1.1) and (H.1.2) in Appendix H determine the representative values, which are used to reflect the variability, by probability analysis. Because the geotechnical design parts in current specifications of highway design still use permissible stress method. Because the safety factors and partial factors in geotechnical design do not have unified values, the *Specifications* retains some parts of permissible stress method.

The *Specifications* gives the design formulas in carrying capacity limit state according to the features of load effects combination and related specifications. Design formulas in serviceability limit state are not given in the *Specifications* because this calculation only suitable for reinforced concrete retaining walls. It may be seen *in Code for Design of Highway Reinforced Concrete and Restressed Concrete Bridges and Culverts* (JTG D62).

Contents about structure importance coefficient can be seen in *Unified Standard for Reliability Design of Engineering Structures* (GB/T 50,283). The safety levels of structures in highway structure design shall be classified into first grade, second grade and third grade according to the importance of the results of structure failures or the technical requirements of structures. The structure importance coefficients of each level are 1.1, 1.0 and 0.9 (approximately half of the reliability index of each level). The coefficients are allow to be adjusted less than one level if it is needed. The results of structure failures of high retaining walls are different from common retaining walls. Table H.1 in appendix use highway classification and height of the wall as the index to determine the importance coefficients. It adjusts the coefficient by half level in the principal that the difference between the origin coefficient and adjusted coefficient is less than 1 level.

Table H.2 in appendix is made according to the contents about 'the impacts on structures' in *Unified Standard for Reliability Design of Engineering Structures* (GB/T 50,283). Loads on retaining walls can be classified into permanent load, variable load and accidental load in accordance with the load time distribution.

The *Specifications* classify the load combinations on retaining walls into type I, II, III according to the features. All the three types of combinations are belong to the common load combinations, which means that it is combined of some permanent loads or combined of permanent loads and variable loads.

The construction load has not been defined clearly in *Unified Standard for Reliability Design of Engineering Structures* (GB/T 50,283). *General Code for Design of Highway Bridges and Culverts* (JTG D60) classifies the constructors and construction equipment into the accidental loads but don't classifies the construction loads. *Unified Standard for Reliability Design of Railway Engineering Structures* (GB 50,216) defines the gravity of some parts of the structures in construction stage and the installation loads as variable loads. The *Specifications* uses the definition in *Unified Standard for Reliability Design of Railway Engineering Structures* (GB 50,216) that classify the construction load into the variable load.

According to the traditional experience in China, earth pressure on the back of retaining walls is usually considered as the active earth pressure and it is usually calculated based on Coulomb's theory. However, in the earth pressure calculation of reinforced retaining wall, the wall needs to be divided by height into static earth pressure section and active earth pressure section.

The *Specifications* treat additional side pressure resulted from vehicle loads by translating the strength of additional loads into the soil thickness. The strength of the additional loads is only determined by the height of retaining wall. The variable load is not divided into calculated load, checking load. The vehicle level is not classified too.

The comparison of the values in earth pressure calculation between some foreign specifications and the *Specifications* is shown in Table 5.1.

To simplify the calculation, the vertical load from vehicles and people can be seen as vertical permanent load. The coefficient may use the partial factor γ_G of vertical permanent load.

The foundation calculation design method in this specification is still based on permissible bearing capacity method but only use the form and terms of limit state

Table 5.1 Values in earth pressure calculation in some specifications

Name of the specifications or standards	Load coefficients		
	Variable load or translated additional earth load	Active earth pressure of filled soil	
		Vertical pressure	Horizontal pressure
American Association of State Highways and Transportation Officials Load and Resistance Factor Design	0.75–1.5	1.0–1.35	0.9–1.5
BS 5400 British Standard. *Steel, concrete and composite bridges. Section 2. Specification for loads*	1.5	(1.0)1.5	(1.0)1.5
The Soviet Union. *Code for Design of Bridges and Culverts for Highways, Railways and Urban Roads*	1.2	(0.7)1.4	(0.7)1.4
Japan. *Example of New Codes for Concrete Structures under Limit State Method*	1.2	1.2	1.2
The *Specifications*	0.95–1.4	0.95–1.4	0.95–1.4

Note If increasing earth pressure has advantage in the structure, the coefficient is the smaller one. If it does not have advantage in the structure, the coefficient is the bigger one

design formula. The two methods are combined through structure calculation of the wall.

The *Specifications* retains the total safety factor method in *Specifications for Design of Highway Subgrades* (JTJ 031—86) Section 3 'Retaining wall stability checking' and adds rules of bearing capacity limit state stability checking equation. Therefore, the results from limit state stability checking formula can be adjusted by the experience of the total safety factor method as the experience prepared for the specification using limit state method in the future.

5.4.3 The Table 5.4 requirements for the embedment of foundation under sloped ground surface is made according to the application experience of retaining walls in railway and highway constructions. 'The horizontal distance from the ground surface' in the table presents the maximum and minimum value of each kind of foundation soil. The value may be chosen according to the geological condition, slope grade and steepness.

To the buried depth of foundation of cutting retaining walls, if the side ditch of a cutting is too deep, the bottom of the side ditch masonry maybe below the foundation of the retaining wall. If seepage happens in the side ditch, the foundation will be impacted. Therefore, this specification stipulate the foundation of retaining wall shall be below the bottom of the side ditch masonry more than 0.2 m.

5.4.4 It is very important for the stability to set drainage holes in the body of retaining walls and set filter layer on the back of retaining walls. In many constructions, drainage hole and filter layer do not have enough attention. The drainage holes can be clogged after 2–3 years use. The groundwater cannot be drain out of the back of retaining walls. This decreases the strength of filling soil on the back of retaining walls and leads to the inhomogeneous deformation on the foundation even failure of retaining walls. Therefore, the design of drainage holes and filter layer of retaining walls need to be pay attention to.

5.4.5 Strength safety needs to be pay attention to when design high retaining wall. Because it needs large amount of rubble, it is difficulty to ensure the qualities of blocks, mortar and the wall. The *Specifications* suggests to use rubble concrete to ensure the quality.

5.4.6 Groin, which is a new developed type of retaining structure, belongs to the gravity block stone structure. Groin is combined with honeycomb cage and rubble or block stones by galvanized steel wire or galvanized aluminum alloy wire. The cage is usually made of corrosion resistant galvanized steel wire, galvanized iron wire and ordinary wire. Groin has advantages of good integrity, flexibility, permeability, deformability and anti-erosion ability. It is also easy to plant and build landscape on groin. Groin has overcome the shortcomings in foundation, groundwater and environment. Groin is suitable for constructions to protect slopes or river banks.

The fundamental principle of groin is the same as masonry gravity wall. The stability under earth pressure is supported by the gravity of wall. The calculation of groin uses the method in gravity wall design.

5.4.7 Rules about cantilever retaining walls in the *Specifications* is made of relevant rules in *Standard for Constructional Quality Acceptance of Railway Subgrade Engineering* (TB 10,025), *Design Manual of Railway Engineering*, *Design Manual of Retaining Structures* and Japanese Code for Design of High-class Highway. All the structures suggested in these specifications are integral pouring structure. There is another type of cantilever retaining wall which is assembled by prefabricated parts. However, this type of cantilever retaining wall has some limitations. It cannot be used in areas with bad geological conditions or seismic regions where seismic intensity is over 8. It also needs lifting equipment and the yard for prefabrication. Its calculation method is different from the method of integral pouring structure. This type of cantilever retaining wall is seldom used in China so that this specification does not have relevant contents.

5.4.8 Anchored retaining wall can be classified into two types that retaining wall with rib columns and plate type of anchored retaining wall. The two types of anchored retaining walls have similar retaining principles but different characteristics.

Anchored retaining wall with rib columns is made of rib columns and vertical wall plates. The distance between anchor arms is bigger than that of plate type of anchored retaining wall. The diameter of anchor arms is usually 100–150 mm. Anchor arms and the hole wall will become a whole after concrete pouring. The anchor arms belong to the type which uses adhesive power as anchor effect.

Plater type of anchored retaining wall is made of the poured integral wall or prefabricated wall with multi-row anchor arms. The distance between anchor arms is bigger than that of retaining wall with rib columns. The diameter of rib columns is usually 35–50 mm. The depth of anchor hole is usually 4–5 m. The mainly used type of anchor arms is slit wedge type. Because the anchor arm contact anchor hole directly, fiction force between anchor arm and hole increases obviously. This type of anchor arm has the characteristics of both sticking type and mechanical type.

5.4.9 Anchored bulkhead retaining wall can be classified into two types that anchored bulkhead retaining wall with rib columns and plate type of anchored bulkhead retaining wall. Anchored bulkhead retaining wall with rib columns is made of rib columns and vertical wall plates. The rib columns are usually set in two rows. This type of anchored bulkhead retaining wall has big anchored area and long anchor arms. It can be used near shoulder or embankment because its deformation is small. Plate type of anchored bulkhead retaining wall has reinforced concrete surface plate. Good appearance can be made by combination of different shapes and thicknesses of surface plate. Plate type of anchored bulkhead retaining wall is mainly used in retaining constructions which need to fit the landscape.

Even though the earth pressure on the plate or wall of anchored bulkhead retaining wall is caused by the additional load from the filler and vehicles but the mechanism of the earth pressure is very complex because the anchored bulkhead retaining wall is a composite structure which has complex interaction between rib columns, anchor arms and the soil. The mechanism is related to the filler properties, compaction, buried depth and length of the rib columns, area of anchor plates and other factors. Railway officials performed a multitude of experiment and reach the following conclusions:

(1) The measured active earth pressure is bigger than the earth pressure calculated through Coulomb's Theory. The ratio is usually between 1.21–1.55. The measured active earth pressure is usually between the calculated active earth pressure and the static earth pressure.
(2) The measured active earth pressure on the back of wall is not in triangular distribution but single peak or zigzag distribution.

5.4.10, 5.4.11 Reinforced earth retaining walls can be classified into reinforced earth retaining wall with panel and reinforced earth retaining wall without panel. Reinforced earth retaining wall without panel is a new developed kind of flexible reinforced earth retaining wall which can fit the foundation deformation well. It can keep the stability of the foundation by confining the lateral deformation of soil mass through the anchor effect of reverse package geogrid. Because the time from reinforced earth retaining wall without panel being applied in highway constructions is not long, further engineering validation is required.

The differences between reinforced earth retaining wall without panel and reinforced embankment are as follow. When the angle between reinforced slop and horizontal plane is more than 70°, the wall is classified among reinforced earth retaining wall. The design of retaining wall needs to follow Appendix H in the *Specifications*. When the angle is less than 70°, the wall is classified among reinforced embankment. The design and calculation comply with *Highway Earthwork Synthetic Material Application Technical Specifications* (JTG/T D32).

The characteristic of materials of tension reinforcement is important in reinforced retaining wall design. The long-term strength and deformation ability of tension reinforcement are related to the stability of retaining wall. The durability is related to the service life of retaining wall. Therefore, choosing suitable material of tension reinforcement is an important process. Toughness and flexibility are needed to avoid embrittlement. The creep, which can result in stress relaxation, must be small to avoid deformation or failure of retaining wall. Therefore, if geogrid is used as tension reinforcement, it needs to has high tensile strength, small ductility and creep, big fiction force with soil and enough corrosion and aging resistance. According to the experience, this specification suggests to use high-density polyethylene (HDPE) geogrid or Polyethylene terephthalate (PET) geogrid in reinforced earth retaining wall without panel.

5.4.12 According to the experience, pile-plank retaining wall is a good type of retaining structure. There was a high pile-plank retaining wall even has 15-m cantilevers. The construction accidents of excessive displacement in pipe or pipe breaking are mainly caused by excessive long cantilevers or increasing rock pressure. The length of cantilever needs to be controlled to keep safety.

The stiffness of anchor piles is greatly different from that of anchor arms. The deformation of anchor arm has obviously impacts on the internal force in pile. Therefore, the deformation of anchor arm needs to fit that of anchor pile.

5.5 Slope Anchorage

5.5.1 The slope anchor technology, which belongs to concealed work, is a developing retaining measure with complex construction processes, different types and many limitations. To keep the safety, the feasibility demonstration needs to be conducted to choose suitable slope anchor style according to the topographic and geographic conditions when slope anchor design.

5.5.2 Prestressed anchor arm is widely used in soil anchor and rock anchor, and the soil anchor layer is mainly sandy soil layer. To ensure the safety and reliability of the slope anchor project, the provisions limit the scope of using prestressed anchor arm, that is, the anchor section of the anchor arm is placed in stable rock layer.

Prestressed anchor arms can be classified into tension type, pressure type, tensile dispersive type and compression dispersive type.

The main characteristic of tensile anchor arm is that the tensile on paste can be translated to surrounding stratums. This type of anchor arm has easy structure and is easy to construct. It is the most widely used type of anchor arm. It is effective especially in soil and hard or medium hard rocks.

The main characteristic of pressure anchor arm is that the pressure is loaded on the paste through bearing body. This kind of anchor arm has good corrosion resistance. However, pressure type anchor arm cannot bear high pressure because the bearing area of the concrete paste is limited by diameter of the hole.

Tensile dispersive and compression dispersive anchor arms can fully use the inherent strength of stratums. The bearing capacity is in direct proportion to the anchored length. Compression dispersive anchor arm entirely uses unbonded steel stranded wire. It has good corrosion resistance because the concrete paste is in compression when compression dispersive anchor arm works. Compression dispersive anchor arm is promoted to be widely used in soil or soft broken rock mass.

5.5.3 Considering the influence of various complex factors, there is a safety margin the design of anchor structure when design slope anchor. Therefore, the safety factor of anchor slope need to be the same as that of the slope without anchors.

5.5.4 It is not needed to consider the safety factor of slope when calculate the anchorage force because the calculation of sliding force E has considered each kind of load and related factors and a safety factor of 1.05–1.30 according to Clause 3.7.7 in the *Specifications.*

The anchorage direction is an important problem in prestressed anchor slope. The most effective anchorage direction is opposite to the sliding direction. However, the construction conditions and sliding boundary conditions limit the anchorage direction. The best anchorage direction needs to be chosen through general comparison to reach the best reinforcement effect. When the dip angle of the sliding surface where the anchor and sliding surface intersect α equals to the Angle between the anchor and the horizontal plane β, the anchor arm has the biggest anti-sliding force but longest length which is uneconomic. There has been some empirical equations of the best anchorage direction. The *Specifications* does not contain these equations because the applicability still need to be tested.

5.5.5 Prestressed anchor is a kind of post-tensioned prestressed component. The limitations of control stress σ_{con} for prestressing of prestressed tendon, especially steel wire, is obviously lower than that of the overground prestressed reinforced concrete structures. The reason is that prestressed anchor is buried in the stratums where has bad work conditions and high risk of erosion. There are many foreign examples that anchor breaks while the control stress of prestressed tendon is more than 0.6 F_{ptk}. Choosing lower control stress σ_{con} of prestressed tendon can also help decreasing the loss of prestressed force in anchor arm.

5.5.6 According to the highway classification and the service life of anchor arm, the *Specifications* revision divides the safety factor of prestressed anchor arm into the cross section design safety factor and bearing safety factor. To the design safety factor value of the cross section of anchor arm, *Specifications for Bolt-shotcrete Support* (GB 50,086—2001) uses 1.6 for temporary anchor arm and 1.8 for permanent anchor arm. *Technical Code for Building Slope Engineering* (GB 50,330—2002) uses the value of 1.6–2.3. Railway specifications use the value of 1.7–2.2. Hydropower code uses the value of 1.8–2.1, which is close to the value used in the *Specifications.*

5.5.7 Rebar which is used in prestressed concrete has been widely used in highway constructions in China. It is used to be called 'strength finish rolled rebar' in China. *Screw-thread Steel Bars for The Prestressing of Concrete* (GB/T 20,065—2006) was presented in 2006 so that the *Specifications* change the name of 'strength finish rolled rebar' into 'rebar which is used in prestressed concrete'.

5.5.8 The protection level and measures is determined by the design service life and the corrosion situation in the environment. The basic requirement of anchor arm anti-corrosion in foreign specifications is to keep the chemical stability of anchor arm in the service life.

3 The surveys and statics in corrosion accidents of anchor arms directed by Fédération Internationale de la Précontrainte (FIP) suggest that a large proportion of corrosion is near the anchor head. Using an anchored retaining wall construction in Hong

Kong Special Administrative Region as the example, the anchor head is corroded because the anchor head wrapping was delayed for a long time after prestressing. A test was conducted in the steel wires of the 45 anchor arms. The steel wires which were delayed by 1–8 months lost 2.7% of the diameter. The steel wires which were delayed by 16–36 months even lost 12% of the diameter. Therefore, the *Specifications* stipulates that the exterior reinforcement bars, anchor arms and bearing plates shall be conducted anti-corrosion protecting measures in time. When using concrete to protect the exterior reinforcement bars, anchor arms and bearing plates, if the thickness of concrete is less than 50 mm, the anchor head is easy to be corroded because the concrete layer is easy to have cracks that can be seeped into by water. Surveys in an anchored slope construction in the southwest China found that the thicknesses of the concrete out of anchor heads are only 20–30 mm. All the steel wires, anchors and bearing plates had serious corrosion. Therefore, the *Specifications* stipulates that the anchor head shall be protected in concrete with the thickness not less than 50 mm.

5.5.10 The slope structure in the *Specifications* only means the bearing pressure structures of anchored slope. The retaining slope structures, such as pile and wall, are not included in it.

Table 5.11 is determined by general considering the factors of construction difficulty, retaining effectiveness, landscape and so on. The type of anchored slope shall be chosen according to the geological conditions and applicable conditions to reduce the loss of prestressed force.

5.5.12 The prestressed anchorage test consists of a basic test and an acceptance test. The basic test needs to be conducted before the construction of anchor arms. The tested anchor arms do not need to work. The basic test is a kind of destructive test used to test whether the anchor design can meet the requirement through the maximum tension can be beard by anchor arms and the safety level of the anchor construction. The acceptance test is a kind of nondestructive test should be conducted in representative working anchor arms after the construction of anchor arms. The test is to check whether the construction quality, including the length of anchor arm, prestressed force and the cohesion strength between the paste and surrounding rock mass, can meet the design requirements. *Specifications for Bolt-shotcrete Support* (GB 50,086) gives the details of the two tests.

In-situ monitoring is very important in anchored slope constructions. In-situ monitoring can be divided into construction monitoring and operation monitoring. The slope anchor is a part of the slope construction. Therefore, the slope monitoring needs to be combined with anchor monitoring. Recently years, the nondestructive testing technology has been developed and applied successfully in constructions. Relevant specifications have been presented in China.

5.6 Soil nailing

5.6.1 Soil nailing is a new technology of retaining slope developed from the new Austrian tunneling method. It is usually used for permanent retaining. Recently years, soil nailing has been developed as permanent retaining structure. Soil nailing mainly bears the tension caused by soil deformation through the cohesion and friction force result from the interaction between soil nailing, bearing plates and rock mass.

Table 5.2 The ratio between the length of soil nailing and the height of wall

Rock classification in slope	Permanent soil nailing retaining wall	Temporary soil nailing retaining wall
Plastic clay	1.2	0.7
Ordinary sand, Clay	0.8–1.2	0.5–0.7
Dense sand soil, Stiff plastic clay	0.8	0.5
Rock	0.6–0.8	–

Highway slope should be permanent construction. The common accidents in soil nailing slopes include that slope cracking results from soil nailing deformation in sections with bad soil condition, bearing plate sink and shear failure of soil nailing result from softened bottom of slope and shortage of bearing capacity in sections containing soft soil and well-developed groundwater, sliding results from the shortage of soil nailing length or anchorage force and the sliding along the rock surface. To ensure the safety of substructure, the *Specifications* limits the application range of soil nailing and stipulates that soil nailing should not be used in corrosion stratums, sections with expansive soil, soft clay, loosened soil and slopes with disadvantage structure surfaces and well-developed groundwater.

Soil nailing is easy to be impacted by water. With the rate of water content increasing in soil, the weight of soil will increase but the shear strength of soil and the surface cohesion between soil nailing and the soil mass will decrease. Most of the accidents of soil nailing constructions is caused by water effects.

5.6.2 The structures of soil nailing include:

(1) The retaining structure made of soil nailing and shotcrete surface with steel grid or geogrid in it.
(2) The retaining structure made of soil nailing, connected with each other by reinforced concrete grid beams, and boundary beams or grade beams. This structure can also be classified into two types by whether it has shotcrete layer under the grid beams.
(3) The retaining structure made of unit soil nailing and nailing headed that was protected by concrete blocks.
(4) The retaining structure made of soil nailing and retaining plates.
(5) The retaining structure made of the continuous plane of cast-in-place or precast concrete plates and soil nailing.

The length of soil nailing is usually 1.2–1.5 times of the height of retaining wall. The ratio may be chosen according to Table 5.2 through the type of soil nailing retaining structure and rock classification.

The distance between soil nails shall be 0.75–3 m. When use hole grouting soil nails, the distance between soil nails may be 10–20 times of the diameter of holes. The density of soil nails on the retaining surface is usually one nailing per 6 m^2. The distance between soil nails can be chosen according to Table 5.3 through the type of soil nailing retaining structure and rock classification.

Table 5.3 The distance between soil nails (m)

Rock classification in slope	Permanent soil nailing retaining wall	Temporary soil nailing retaining wall	
	Hole grouting soil nailing	Hole grouting soil nailing	Struck nails
Sandy soil	1.5	≤1.0	≤3.0
Dry hard clay	2.0		
Rock	≤3.0		

The angle between the soil nailing and horizontal plane should be in the range of 5°–25°. The smaller angle means less deformation of retaining wall. Therefore, the angle may be close to 0 when grouting by pressure with effective exhaust measures. The angle should be more than 15° if the soil nails are grouted by gravity. The angle of top soil nails may be increased so that the ends of nails can be plugged into the deeper hard rocks.

5.6.3 The structure calculation of soil nailing support includes internal and external overall stability checking, support surface course, and calculation of the connection between the surface course and soil nailing.

There are still some problems in the limit state method based on the probability theory. Therefore, the calculation of soil nailing support structure in China still uses general safety factor method. *Guidelines for Soil Nailing in Highway Engineering* still uses general safety factor method in the calculation but it uses the characteristic value of the design value of soil mechanical parameters not the mean value. It means that the characteristic value of the soil limit strength is set as the design value of the soil strength. The design value of earth pressure, used in the analysis of the overall stability and the sliding of soil nailing support, and the design value of the lateral earth pressure, used for calculating the internal force in soil nailing, are all characteristic values. To design the concrete components such as surface and connectors [according to the *Code for Design of Concrete Structures* (GB 50,010)], a partial safety factor for load need to be multiplied with the earth pressure loading on the surface.

Internal and external overall stability checking, support surface course, and calculation of the connection between the surface course and soil nailing need to be conducted according to the rules in *Guidelines for Soil Nailing in Highway Engineering*.

5.6.4 In-situ test of soil nailing is very important in soil nailing support constructions. Many accidents of soil nailing support constructions happened because the tests were not conducted so that the mistakes in design decreased the construction quality. The in-situ test of soil nailing needs to be conducted by the test and detection unit which is entrusted by the owner.

Generally, the test does not conduct uplift test on the working piles to determine the ultimate load in nondestructive test method. The cohesion length is too long if a tension is on the nailing head. This is different from the real condition because soil nailing is easier to yield by tension in this condition. The maximum tension in nailing head in test also cannot be directly compared with the internal force of working soil nailing to determine whether the anti-lifting ability meets the requirement because their cohesion lengths are not the same.

The requirement of the soil condition in soil nailing creep test in Chinese specification is $I_p > 17$. It is $I_p > 20$ in French specification. *Guidelines for Soil Nailing in Highway Engineering* uses the rule of $I_p > 20$ and stipulates that $\omega_L > 50\%$. According to the plastic graph classification, the equation of line A is that $I_p = 0.73(\omega_L\text{-}20)$.$I_p = 22$ when $\omega_p = 50\%$. The soil with $I_p = 20$ and $\omega_p = 50\%$ is located at the above of line A and the right of line B. It means that the clay with high liquid limit and plasticity index needs soil nailing creep test.

5.7 Slide resistance pile

5.7.1 Construction experience suggests that the main causes of the failure of anti-slide piles include: shortage of geographic conditions, inaccurate geological parameters, unclear boundary of sliding area, inaccurate judgement of the position of the sliding surface, unclear information of each sliding surface in the area where happened times of landslide, differences between tested strength of rock in sliding surface and the real condition, loss of the consideration on the slope stability above the pile. The prospecting and design of anti-slide pile must be combined together as a system. Detailed geological data is the key point of design. To avoid failures result from inaccurate geological data, the results of geological prospecting need to be checked in time according to the monitoring information of geological conditions and slope deformation in the excavation process.

5.7.3 The retaining structure combined of anti-slide piles and prestressed cables is usually used in rock sliding bed. The anchorage segment shall be set in stable rocks.

The stiffness of anchor pile is obviously different from that of anchor arm. The deformation of anchor cable will obviously impact the internal force in pile in the design of cable anti-slide pile. Therefore, the deformation of cable has to be controlled to meet pile's deformation. Otherwise the design value of the internal force must be greatly different from the real value, for example the design value is a sagging moment but the real value is a negative moment.

5.7.4 According to the requirement of durability design, the *Specifications* increases the requirement of strength grade of concrete from C20 to C30.

To avoid big deformation of steel framework in the producing and assembling process, *Code for Design of Ground Base and Foundation of Highway Bridges and Culverts* (JTG D63—2007) and *Code for Design on Subsoil of railway Bridge and Culvert* (TB 10,002.5—2005) stipulate that the minimum diameter of main reinforcement shall be 16 mm and the amount of main reinforcement shall be more than 8. Anti-slide pile is mainly used to bear horizontal loads so that the amount of main reinforcement is usually more than 8.

When hole and pour concrete, the compaction usually relies on the gravity of concrete itself because tamping is difficult. To ensure that concrete can overflow steel framework smoothly, the main reinforcement should not be arranged too dense. The *Specifications* suggests that the clear spacing should not be less than 120 mm.

Table 5.4 Foundation coefficient of anti-slide pile (the ratio increasing with depth)

No	Soil name	Vertical direction m_0 (kPa/m^2)	Horizontal direction m (kPa/m^2)
1	Soft plastic clay and silty clay ($0.75 < I_L < 1.0$), Silt	1,000–2,000	500–1,400
2	Soft plastic clay and silty clay ($0.5 < I_L < 0.75$)	2,000–4,000	1,000–2,800
3	Stiff plastic clay and silty clay, Fine sand and medium sand	4,000–6,000	2,000–4,200
4	Stiff clay and silty clay, Coarse sand	6,000–10,000	3,000–7,000
5	Gravelly sand, detritus soil and pebble soil	10,000–20,000	5,000–14,000
6	Dense boulder	80,000–120,000	40,000–84,000

Note 1. I_L is the liquidity index. Its foundation coefficients m_0 and m correspond to the displacement of 6–10 mm on the pile top
2. The value choosing is not limited by this table if there is credible experience

In difficult circumstances, the clear spacing can be reduced but shall not be less than 80 mm.

According to the requirement of durability design in *Code for Durability Design of Concrete Structures* (GB/T 50,476—2008) and *Specification for Deterioration Prevention of Highway Concrete Structures* (JTG/T B07-01—2006), the thickness of concrete cover for reinforcement (including main reinforcement, stirrup and distributed reinforcement) shall not be less than 70 mm.

5.7.5 The displacement and internal force of the pile below the sliding surface need to be calculated as elastic foundation beam. The calculation method needs to be chosen according to the distribution of foundation coefficient. When the coefficient distributes in triangle, the displacement and internal force is calculated in 'm' method. When the foundation coefficient is a constant, the displacement and internal force is calculated in 'K' method. The foundation coefficient below the sliding surface is determined by stratum property. Clause 10.2.8 of *Code for Design on Retaining Structures of Railway Subgrade* (TB 10,025—2006) suggests that the foundation coefficient of complete rock stratums or stiff clay should be a content K. The foundation coefficient of stiff plastic-half dry sandy clay and gravels or weathered detritus soil usually distribute in triangle when there are not slide mass and overload on the sliding surface. Otherwise the coefficient distributes in the shape of trapezoid.

Choosing suitable foundation coefficient is very important in the calculation. But its test is very difficult. Table 5.4 and 5.5 give the foundation coefficients of anti-slide pile and other stratum mechanical parameters according to *Code for Design on Retaining Structures of Railway Subgrade* (TB 10,025—2006).

5 Transverse allowable bearing capacity $[\delta_H]$ of the subgrade surrounding pile in different topographical and geological conditions may be calculated in the method as follow:

Table 5.5 Foundation coefficients of anti-slide pile and stratum mechanical parameters

Stratum classification	Internal friction angle (°)	Modulus of elasticity E_0(kPa)	Poisson ratio μ	Foundation coefficient K(kPa/m)	Shearing stress (kPa)
Fine-grained granite, Syenite	>80	5,430–6,900	0.25–0.30	2.0×10^6–2.5×10^6	>1,500
Diabase, Porphyrite		6,700–7,870	0.28	2.5×10^6	
Middle grained granite	>80	5,430–6,500	0.25	1.8×10^6–2.0×10^6	>1,500
Coarse-grained Syenite, Hard dolomite		6,560–7,000	0.25		
Hard limestone	80	4,400–10,000	0.25–0.30	1.2×10^6–2.0×10^6	1,500
Hard sandstone, Marble		4,660–5,430			
Coarse-grained granite, Granitic gneiss		5,430–6,000			
Medium hard limestone	75–80	4,400–9,000	0.25–0.30	0.8×10^6–1.2×10^6	1,200–1,400
Medium hard sandstone		4,660–5,000			
Medium soft granite		5,430–6,000			
Hard shale	70–75	2,000–5,500	0.15–0.30	0.4×10^6–0.8×10^6	700–1,200
Common limestone		4,400–8,000	0.25–0.30		
Common sandstone		4,600–5,000	0.25–0.30		
Hard marlite	70	800–1,200	0.29–0.38	0.3×10^6–0.4×10^6	500–700
Medium hard shale		1,980–3,600	0.25–0.30		
Medium soft limestone		4,400–6,000	0.25–0.30		
Medium soft sandstone		1,000–2,780	0.25–0.30		
Medium hard marlite	65	700–900	0.29–0.38	0.2×10^6–0.3×10^6	300–500
Common shale		1,900–3,000	0.15–0.20		
Soft limestone		4,400–5,000	0.25		
Medium soft marlite	45	30–500	0.29–0.38	0.06×10^6–0.12×10^6	150–300
Hardened clay		10–300	0.30–0.37		

(continued)

Table 5.5 (continued)

Stratum classification	Internal friction angle (°)	Modulus of elasticity E_0(kPa)	Poisson ratio μ	Foundation coefficient K(kPa/m)	Shearing stress (kPa)
Soft schist		500–700	0.15–0.18		
Hard coal		50–300	0.30–0.40		
Dense clay	30–45	10–300	0.30–0.37	0.03×10^6–0.06×10^6	100–150
Common coal		50–300	0.30–0.40		
Cemented pebble		50–100	–		
Soil mixed with rocks		50–100	–		

(1) If the stratum is rock stratum, it can be calculated by Eq. (5.1):

$$[\delta_H] = K_H \eta R_c \tag{5.1}$$

where:

$[\delta_H]$—transverse allowable bearing capacity of subgrade;

K_H—the conversion factor in horizontal direction. It ranges from 0.5–1.0 according to the integrity of rock, stratification, interlayer cement and the cementation condition, density of joint and filler;

η—reduction factor which ranges from 0.3–0.45 according to the fracture development and weathering degree;

R_c—uniaxial compressive ultimate strength of rock (kPa).

(2) If the stratum is made of soil or weathered soil or sandstone and conglomerate, it may be calculated by Eq. (5.2) when the cross slope i is small and Eq. (5.3) when i is big and $i \le \varphi_0$.

$$[\delta_H] = \frac{4}{\cos\varphi}[(\gamma_1 h_1 + \gamma_2 y)\tan\varphi + c] \tag{5.2}$$

$$[\delta_H] = 4(\gamma_1 h_1 + \gamma_2 y)\frac{\cos^2 i\sqrt{\cos^2 i - \cos^2\varphi_0}}{\cos^2\varphi_0} \tag{5.3}$$

where:

γ_1—unit weight of the soil mass above sliding surface (kN/m^3);

γ_2—unit weight of the soil mass below sliding surface (kN/m^3);

φ—internal friction angle of the soil mass below sliding surface (°);

c—cohesion strength of the soil mass below sliding surface (kPa);

h_1—the distance between sliding surface and the ground (m);

y—the distance between sliding surface and calculating point (m);

φ_0—the general internal friction angle of the soil mass below sliding surface (°).

6. Substructure Widening and Reconstruction

6.1 General requirements

6.1.1 Before widening and reconstructing substructure, investigations, surveys and tests need to be executed on the existing substructure in fill materials, moisture content, density, compaction, strength, and stability as well as the type and causes of substructure defects. This is the base work to determine utilization and widening plan of the existing substructure. The *Specifications* defines the contents required in surveys and tests of existing substructure.

6.1.2, 6.1.3 Determining utilization and widening plan of the existing substructure is an important part in substructure widening and reconstruction design. There are 3 kinds of utilization plan of the existing substructure. Plan 1 is to use the existing substructure directly. This plan is suitable for the sections where there is no defects. The substructure strength can directly meet the requirement of reconstruction. Plan 2 is to treat the existing substructure before reconstruction. This plan is suitable for the sections where little defects or a shortage of strength exist. The substructure can meet the requirement of reconstruction through treatments. Plan 3 is to remove the existing substructure and reconstruct a new one. This plan is suitable for the sections with serious defects. The substructure cannot meet the requirement even through treatments. The plans need to be chosen after a techno economic appraisal of the condition of the existing substructure and the goals of reconstruction.

According to the different relative positions between widened substructure and the existing substructure, substructure widening measures can be classified into 3 types including split joint type, separate type and mixture type. Each widening measure has its own applicable condition which is shown in the Table 6.1 and Fig. 6.1. Most of the motorway widening constructions in China use the double-side split joint measure. Few of motorways, mainly being long-span bridges, are widened through double-side separate type measure. If a motorway has reserved the space, the widening can be conducted in the median divider. If there is no interchanges in a long section (usually >5 km) of motorway, the widened motorway can be separate from the existing motorway when the land, time and traffic are limited.

6.1.4 The most obvious problem in substructure widening construction is the differential settlement between origin and reconstructed substructures. The loads of widening construction can lead to a new settlement in the origin substructure. The reconstructed substructure also has the settlement resulting from compaction. Creep and sliding can happen between the origin and reconstructed substructures because of low-quality split joint. There are some necessary engineering measures need to be taken to reduce the differential settlement between the new and existing substructure to prevent longitudinal cracks. The new substructure shall maintain a good connection with the existing substructure. Treatments on the new substructure can also decrease the settlement. Compacting on the embankment is important to decrease the settlement in itself.

Table 6.1 Widening type classification

Widening type			Advantages	Disadvantages
Split joint type	Single-side Widening	Figure 6.1a	It only needs little adjustment in alignment. Construction is easy in the widening side	The two-way cross slope need to be changed into one-way cross slope The structure is difficult to treat It is difficult to set interchanges and service facilities. Differential settlement makes split joint difficult. Clearance may not be enough to meet the requirement for down motorways or navigation
	Double-side Widening	Figure 6.1b	It only needs little adjustment in alignment. The traffic composition does not need to change	Differential settlement between old and new substructure makes split joint difficult Clearance may not be enough to meet the requirement for down motorways or navigation
	Median widening	Figure 6.1c	The alignment and traffic composition all do not need changes	The median divider must be wide enough
Separate type	Single-side Widening	Figure 6.1d	It only needs little adjustment in alignment. Construction is easy in the widening side	The two-way cross slope need to be changed into one-way cross slope The structure is difficult to treat The land use is large
	Double-side Widening	Figure 6.1e	The alignment does not need changes	The traffic composition must be changed The land use is large
Mixture type	Double-side split joint type or separate type	Figure 6.1e	It has the both advantages of (b) an (e)	The traffic was separated or combined in the route. Traffic composition is complex. It is difficult to keep safe The construction is difficult in splat joint part. Traffic function is not good in separate part

6.2 Investigation and evaluation of the existing substructure

6.2.1–6.2.4 It is crucial to survey, test and evaluate existing substructure and subsoil before the design of widening existing substructure.

Existing substructure investigation, which is an integrated way operated step by step, is combined of data collection, field survey, and test. First, the survey and

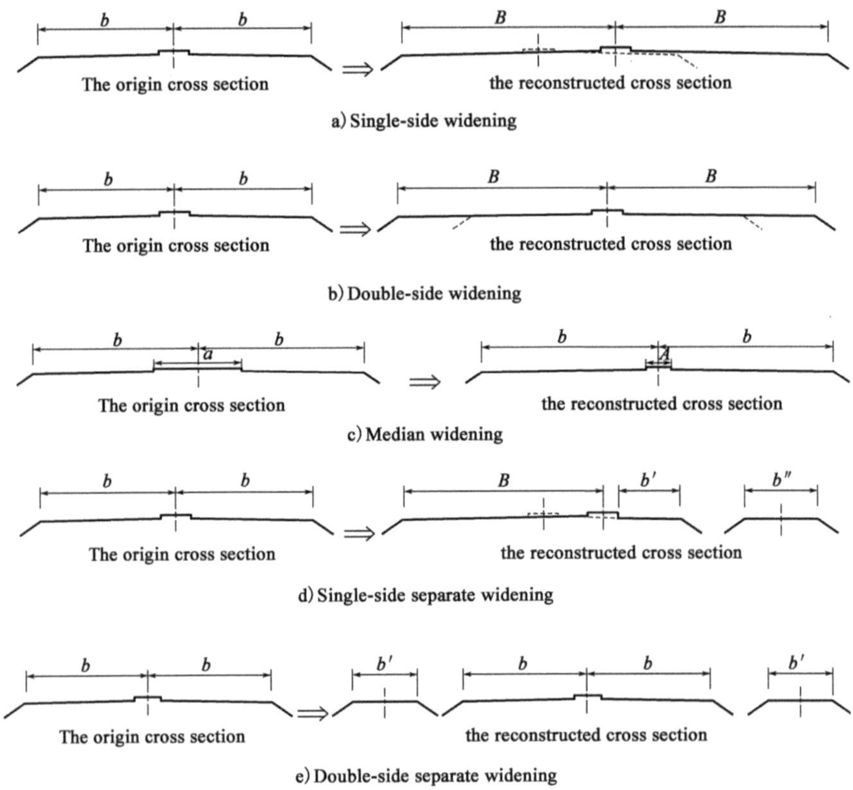

Fig. 6.1 Widening types

design, construction material (including exploration and design file, test report, as-built drawing and so on), maintenance data and monitor data of existing substructure and pavement need to be collected; second, the field investigation should be compared with origin related data to know the existing substructure condition and road segment. Third, make corresponding survey test, test work program and detect existing substructure and pavement disease according to various road section condition.

Investigation including three major part: investigation of existing substructure pavement condition, test of pavement strength and survey of subgrade soil, prospection under the existing substructure and foundational prospection in the widening area. The major objective of prospection of existing substructure is to determine the equilibrium moisture content in substructure and resilient modulus to identify whether the substructure performance is fully utilized or exists any deviation. In general, the degree of compaction, strength, hydrological status, the degree of consolidation and slope stability should be analyzed and compared by distributing protection hole to determine the use of the existing substructure and the substructure widening program which provide evidence for substructure widening. The widening of special substructure need to prospect what will affect the stability of widening substructure and other deformation.

Subgrade dynamic resilient modulus has notable stress dependence. Subgrade modulus, which can be back formulated by dynamic deflection test with FWS, exists discrepancy in dynamic resilient modulus in substructure design on stress level needing amendment. According to domestic and international research results, existing substructure resilient modulus can be considered as the results got by dynamic deflection test with FWS multiply by correction factor from 1/4 (concrete pavement) to 1/3 (asphalt concrete pavement).

6.3 Substructure widening and reconstruction of Class-2 and lower-class highways

6.3.2 The standard of substructure widening and new line reconstruction need to be operated based on corresponding highway technical standards. Where the substructure height is restrained by flood stage shall comply to the requirements specified in Clause 3.3.2; when substructure height is affected by moisture and frost depth, reconstruction may improve the moisture of substructure to avoid adjust height.

6.3.4 During the substructure widening, the diseases like differential settlement and deformation crack usually happened between the new and old subgrades. The major reasons are the different density and characters of fillings, water infiltration and improper connection.

To reduce the differential settlement between the new and existing substructure and prevent substructure diseases shall enhance the fillings of widening substructure, connection with the existing substructure and new substructure and drainage design. The fillings used for widening and reconstruction of embankment shall be matched with the existing substructure fillings. A higher permeability coarse-grained soil is prior to use. When fine-grained soils are used for fillings, subgrade strength, resilient modulus, drainage design in subgrade (especially the connection between the new and existing substructure), necessary horizontal and vertical leak ditch need to meet the standards to drain the water in the subgrade.

To ensure the integrity of widening, the connector needs to be taken by benching which is to clean loose soil on the slope and bench on the existing slope from the bottom to the top step by step. A geogrid need to be laid horizontally for an embankment with a higher height to enhance the whole stability of the substructure and reduce the differential settlement.

Engineering experiences show widening on the embankment slope directly is forbidden. Some engineering measures, such as filling the substructure with an extra width or scarifying the existing substructure, may be adopted when the widening width is less than 0.75 m to ensure the compaction of the widening substructure.

6.4 Widening and reconstruction of existing substructure for Motorways or Class-1 highways

6.4.3 The substructure of highway sections on soft soil has almost consolidated into the settlement stable state after a long-term operation. However the foundation near widened substructure is still in the origin state. The load of new substructure will lead to the additional settlement, which will influence the existing substructure, on the foundation. Therefore, the key point of the widening and reconstruction design of the substructure of sections on soft soil is the control of differential settlement. The old version of the *Specifications* (JTG D30—2004) stipulated that the differential settlement between the new and old substructures shall be controlled during the substructure widening and the post-construction increment of the crown cross slope of the existing and widening substructures shall not be greater than 0.5%, according to the relevant researches and construction experience in Jiangsu, Zhejiang and Guangdong.

The soft soil foundation treatment design of new-constructed substructure uses the post-construction settlement as the control index of settlement. However it is different in the soft soil foundation treatment design of widened substructure. The design needs to consider the condition and developing regulation of the existing substructure and the differential settlement. It uses the 'maximum post-construction settlement of crown slope' as the settlement control index. Some constructions use the control standard that the post-construction settlement should not be more than 100 mm and treat soft soil foundations with CFG piles or rigid piles. These constructions do not pay attention to the settlement of the existing substructure. As the result, there is only settlement on the existing substructure leading to deformation and poor drainage on the pavement. This effects the driving safety. The 10-year experience has prevented the rationality of the control standard in the old version of this *Specifications* (JTG D30—2004). The *Specifications* continues to use the relevant contents.

Because the construction settlement in the refilling process of substructure widening will directly impact the settlement of the existing foundation, the drainage consolidation method, which is aimed to decrease post-construction settlement by increasing the consolidation speed but ignores the total settlement, no longer applies. Dynamic compaction is not suggested in this specification because it will impacts the settlement and stability of the existing substructure. When the widened substructure is in the sections near fishpond, river or reservoir, anti-seepage measures need to be conducted before drainage to avoid cracks result from additional settlement caused by drainage. There has been examples of drainage resulting in excess deformation even collapse on substructure in the real constructions.

6.4.7 According to the surveys in water content of highway subgrade soil, the water content in subgrade soil is generally higher than the water content at time of completion. There is also an obviously decreasing in resilient modulus and compaction. Using the resilient modulus standard of new-constructed substructure, a large scale of excavation and treatment is needed. Therefore, well-directed treatments combined with pavement utilization and widening design may be conducted according to construction features and the causes and impacts of substructure defects in the principles of resource conservation and utilizing the existing structures.

7. Special Substructure

7.1 General requirements

7.1.1 Special substructure includes special soil (rock) substructure, poor geological substructure, and special designed substructure which is severely affected by natural factors such as water and climate. Special geotechnical soils include soft soil, red clay, high liquid limit soil, expansive soil, loess, saline soil, permafrost, desert and so on. Poor geological phenomena include landslide, collapse, rock pile, debris flow, karst, mined-out area and so on. Substructure under special conditions refers to the substructure which is affected by natural factors such as water or climate, including snow damage, saliva flow ice, coastal, reservoir substructure and seasonal permafrost.

Geotechnical engineering properties differ greatly with different special soil (rock), poor geological phenomena and special climatic conditions. Also, the main factors affecting the long-term performance of the substructure, substructure disease, and the extent of road hazards is not the same. At the same time, it is very important to carry out the in-situ testing of special geotechnical properties because of that the special geotechnical soils are greatly influenced by environmental condition, especially sensitive to the water environment. In addition, the indoor experiment is difficult to reflect its actual engineering properties. Therefore, in the process of substructure design, it needs to carry out comprehensive geological survey for the particularity of these geological bodies, to find out the nature, cause type, scale, stability, development trend, and the degree of damage to the highway of special geological body. These can provide the reliably geological basis for the substructure design.

7.1.2 The areas existing large-scale special geological body with complex nature are easy to induce geological disasters, endanger the safety of highway directly, and result in a large remediation projects with high cost and difficulties to cure the road disease, which bring security risks to road operators. Therefore, for sections affected by special soil (rock), poor geology, water or climate and other natural factors, the geological line selection work is very important. When designing the geological line, it is needed to avoid large-scale, complex nature, poor geology difficult to deal with, and special soil (rock).

7.1.3 Climatic environment, water, geology, and other factors have profound influence on the long-term performance of special substructure. If improper engineering measures are taken, it is easy to cause more serious substructure disease. Therefore, the special substructure design needs to be combined with the prevention and control of substructure disease, follow the principle of prevention in the first place and integrating prevention with control, do well in the construction of substructure structure, filler selection, foundation treatment, drainage and protection and other integrated design, control environment (such as water, temperature, humidity, etc.) changes which may have impacts on the substructure, prevent and control the substructure disease. For the existing substructure disease, the treatment, multi-program technical and economic comparison should be carried out. Effective engineering measures should be taken to strive to cure substructure disease and leave no accident possibility, according to local conditions.

7.2 Substructure in landslide areas

7.2.1 Landslide refers to a kind of geological phenomenon under a certain topographical and geological conditions. Due to various natural or human factors, the unstable rock and soil slides along a certain weak plane on the slope under the action of gravity. Landslide is one of the major diseases of mountain road and dose harm to the mountain road. Therefore, it is necessary to attach immense importance to the investigation of landslide. Through comprehensive investigation, the distribution range, the cause and the nature of the landslide can be identified, the stability degree of the landslide and the dangers of the highway can be judged, and the reliable geological parameters can be achieved for the landslide prevention and control.

For large-scale landslides with complex nature, the main choice is the bypassing because of the large-scale renovation project, the unknown nature of the landslide, and the low reliability engineering. For slow deformation landslides and the ones which are difficult to find out their nature in a brief time, under the premise of ensuring the safety of the route, take the principle of comprehensive planning and phased rectification, construct the emergency project which is conducive to the stable landslide and establish the necessary observation system to observe the effect, master the information of landslide changes, and protect landslide radically step by step.

Landslide formation and development is the result of a variety of factors. Governance should distinguish between primary and secondary factors to carry out the comprehensive control. Water is the primary factor in inducing landslides. It is very important to prevent water from entering the slide and to discharge water out of the landslide body. Load reduction is an effective way to slow down the deformation of the landslide, to ensure the safety during the construction period, and to reduce the retaining projects. Landslide has diverse types, there are different situations even in the same type of landslide. Therefore, it is needed to apply various flexible control techniques and comprehensive management based on the specific analysis of the landslide situation.

Engineering experience shows that when constructing road in the fractured zone, special geotechnical and deep cutting with loose soil, broken soft rock high slope, high cutting slope with negative weak plane, slope with poor embankment, and engineering sections prone to landslides where surface water collects and groundwater is abundant, preventive measures need to be taken to set up necessary reinforcement project to avoid the landslides.

7.2.2 Correct evaluation of landslide stability and analysis of hazards that landslides exert on the highway is the key point in landslide control design. Stability safety factor F_s of the landslide should be comprehensively considered and selected according to the scale of the landslide, the deformation rate and the degree of damage, combined with the development stage of the landslide, the grade of the highway and its importance, as well as the nature of the landslide, the sliding factor, the sliding body, its geotechnical strength index and reliability. Stability safety factor under special condition may be increased or decreased as necessary after demonstration.

Loads that is considered in the calculation of landslide includes permanent load such as gravity of slide body, the additional load caused by the buildings on the landslide, dynamic load (such as vehicle load), the earthquake force and the temporary load caused by construction on the landslide. Although the duration of the temporary load applied to the landslide is short, it is very adverse for the stability of the landslide, so the temporary load needs to be analyzed. When considering the influence of the earthquake on the ancient landslide, investigation and analysis is needed to find out whether the landslide has been influenced under the same effects of the earthquake which is at the same level of the design. When conducting back-calculation of c and φ, the effects of earthquakes in history need to be considered.

The stability analysis of landside uses a lot of mechanical inspection methods; the transmission coefficient provided by codes is the explicit solution formula of imbalance thrust force method. It will be hidden in the shear strength index and safety factor of transmission coefficient. The sliding force is only multiplied by a safety factor with the concept of overload coefficient. The transmission coefficient assumes that the action direction of thrust force T_i is taken as the direction of upper side bar and the vertical safety shear force is introduced. Therefore, the safety factor obtained by the transmission coefficient method is biased towards safety. When the stability coefficient is given, the sliding force T_i calculated by the transmission coefficient method may be used as the thrust of the retaining structure.

The test selects direct shear test method of rock mass as far as possible about shear strength index of sliding surface. The peak intensity is used when the landslide is sliding at the first time; the shear or ring shear is used for many times and the residual strength is used when the sliding is frequent or sliding displacement is large; when there is stagnant water in the sliding zone, quick shear under saturation or quick shear under controlled moisture content is needed; when the sensitivity of sliding zone is high, it is necessary to carry out the test in situ. Large area shear test is required when the content of coarse particles in the materials of sliding zones exceeds 30%.

The anti-inference method for the value of c and φ needs to pay attention to back calculation method, adaptability to geological conditions, the difference between the landside geological conditions and the most unfavorable conditions which may occur later and reasonable analysis for the values of c and φ. Experience data has its specific characteristics and limitation, so it is necessary to pay attention to its using conditions.

7.2.4 The drainage project of landslide surface has a great effect on its stability. When surface drainage is laid out, the flow of surface water should be avoided and the surface water in the sliding range can be eliminated quickly. Subsurface drainage consists of shallow and deep layers. The shallow groundwater is using various forms of sewers commonly and the deep groundwater is using inclined drainage hole and drainage tunnel.

7.2.5 Load reduction plays a key role in slowing down landslide deformation. Load reduction can be used as one of the main means of landslide control and large landslides needs to be combined with other projects. Load reduction has added new exposed surface and the design needs to fully demonstrate whether it will cause secondary landslide or worsen the condition of the original landslide, so it is necessary to choose carefully. Attention should be paid to the stability of counter pressure treatment measures and new landslides should be prevented.

7.2.6 Anti-slide retaining wall is one of the effective measures of political landslide and is often as part of drainage, load reduction and other comprehensive measures and combined with the supporting sewer. It is unfavorable to the stability of the landslide when the excavation depth of anti-slide retaining wall is large. Measures need to be taken to prevent the stability of the landslide during construction such as short excavation, timely construction and so on.

7.2.7 Anti-slide pile is widely used as an effective anti-slide measure for stabilizing landslide. It has some advantages such as flexible layout, easy construction and construction with little influence on landslide stability. The prestressed anchor rope and anti-slide pile make up the anchored slide-resistant pile so as to improve the force of the pile and reduce the length of the pile section and anchor section. The effect is better. Attention needs to be paid to the influence on the stability of the landslide when the pile foundation is excavated during the construction of anti-slide pile.

7.2.8 The technology of prestressed anchor rope has been widely used in the landslide control project. The anchor section is generally in the stable rock stratum and the example of soil layer is less. The design need to take measures to prevent the relaxation of prestressed anchor rope, soil compaction and settlement of the frame, the ground beam and the anchor pier, partial collapse often cause prestress relaxation of anchor rope and the failure of the anchor rope. Therefore, the measures of anti-scouting should be taken for soil slope.

7.2.10 Landslide control monitoring includes construction safety monitoring, control effect monitoring and dynamic long-term monitoring and give priority to construction safety monitoring and control effect monitoring. The result of construction safety monitoring is an important basis for judging the stability of landslide, guiding the construction and checking the effect of prevention and control. Construction safety monitoring includes ground deformation monitoring, surface crack monitoring, deep displacement monitoring of sliding body, groundwater level monitoring, pore water pressure monitoring, in-situ stress monitoring and so on. The monitoring results need to be combined with the construction safety and long-term monitoring in order to understand the characteristics of landslide change after the implementation of the project, and provide a scientific basis for the final acceptance of the project. Generally, the monitoring time of the control effect is not less than one hydrological year after opening to traffic.

7.3 Substructure in rockfall and talus sections

7.3.1 Rockfall and caving-in are collectively called collapse. The falling of small stones on the hillside is often called debris. The collapse referred to in this section is a general term for scattered, collapsed, rockfall, and dangerous rocks. The collapse of highway on mountain area is often encountered in varying degrees, and the damage to highway operation safety caused by collapse cannot be ignored. Therefore, the distribution range, stability and hazard scope of the rockfall need to be found out by means of comprehensive investigation and the route line need to be rationally laid out to avoid possible collapse. Effective control measures can be taken to ensure the safety of highway operation based on stability evaluation and prediction analysis.

7.3.3, 7.3.4 In recent years, the slope anchorage and flexible protection technology have been popularized and applied and good results have been achieved for serious rock fragmentation and slope collapse. The flexible protection system includes active form and passive form. The active protection system is composed of a system bolt and a protective net, and the passive protection system is composed of an intercepting net. In general, active protection systems are preferred.

7.4 Substructure in the section backfilled with rocks

7.4.1 Rock mass is a loose deposit which is from the steep hillside rock collapse material by gravity handling piled up in the hillside toe or gentle hillside. There are some difficulties in determining the accumulation pattern and the accumulation thickness of rock mass. Drilling is difficult to bore and the geophysical signal transfer is weak because the rock mass is relatively loose. Therefore, field investigation and mapping work need to be strengthened in rock mass reconnaissance.

7.4.2 The sliding of ground interface under rock file or weak structural plane can be induced when the rock underlying rock soil interface slope is large or have the weak structural plane coupled with the substructure load or excavation of a part of substructure, this case is quite common. The design shall base on the experimental data analysis, take into account the role of surface water and ground water and use the most unfavorable physical and mechanical indicators for stability analysis.

7.4.4 The substructure of the rock lot is prone to larger settlement under the driving load or earthquake load because of the large pore and loose structure, resulting in the destruction of the pavement structure. Therefore, it is necessary to strengthen the treatment of rock mass below the base of the road surface, it also needs to meet the requirements of settlement deformation in addition to meeting the stability requirements.

Science and technology project of western department of traffic 'The key technology research after the construction of Mianmao highway' conducted a special study on rock substructure stability and settlement control. The projects use three kinds of treatment methods which include filling, dynamic compaction and grouting. The rock mass with relatively small particle size is crushed by relatively large rock blocks and the effect is better by dynamic compaction; the stable rock pile can meet the engineering requirements by using the replacement of geotextile measures; large size rock pile can play a dual control effect of stability and settlement by using the grouting filling.

7.5 Substructure in the section of debris flow

7.5.1 The debris flow is intermittent torrent which carries a large amount of mud and rock intermittent torrent and mainly affected by rainfall, snow and ice melt and is multiple. When the route passes through the debris flow ditch, it is necessary to strengthen the field investigation and residents visit along the gully and find out the geomorphic features along the debris flow ditch, the scale, the material composition, the development trend and the harm degree of the debris flow.

To control debris flow and eliminate debris flow effectively, it is necessary to combine comprehensive measures such as soil control, water control and platoon guidance for the frequent debris flow. But for the comprehensive treatment of debris flow, a non-highway department can undertake and need to coordinate with prevention and control plan of other local departments and need comprehensive planning and co-governance.

7.5.2 The principle of bridge across the debris flow is: set up a bridge rather than culvert, use a large span bridge rather than a small span bridge or porous culvert, it is especially true for viscous debris flows and mountain debris flows. It cannot be determined only by the flow calculation while determining the pore size of a bridge. It needs to be combined with the topographic conditions, the width of the trench, the nature and trend of the debris flow, and the law of its development and change.

There are many adverse conditions of culverts compared with bridges. The main adverse condition is small span, low headroom, gentle discharge longitudinal slope, long flow, large peripheral resistance, poor ability of debris flow drainage, easy catharsis of debris flow blocking and silting, difficult rescue; engineering practice shows that the debris flow rate is far higher than the small bridges and culverts and culvert buried across the serious debris flow. Therefore, the debris flow area needs to be careful with culverts.

The maintenance of the debris flow with large or long flowing water during the rainy season is large. It is difficult to ensure passage and should not be used overflow pavement.

7.5.3 The drainage measures include drain and guide ditch, chutes, aqueducts, and training banks.

The drain and guide ditch is constructed by artificial excavation or fill in the flow section, or use the natural channel, and it is mostly built in the debris flow fan or the accumulation terrace, which causes the debris flow to follow a certain route of excretion. The drainage ditch can be used alone or in combination with the retaining works. The design requirement of drainage ditch is that the cumulative silting and erosion damage of building safety are not present in the years of use.

The aqueduct can be used to discharge debris flow less than 30 m^3/s, which is generally adopted under certain topographic conditions, and the slope is suitable for 8% to 15%. It is not advisable to use or use caution for the volatile flow of the channel, the abrupt change of erosion, the high viscous debris flow and the large boulders containing large particle size of the solid material.

The chutes are used to prevent congestion and blockage of the bridge. Training bank is used to change the flow direction and flow rate of the debris flow, so that the debris flow can be smoothly lined up to ensure the safety of the substructure.

7.5.4 The dam is constructed in the formation zone or form of debris flow—a man-made structure in the circulation area, which aims to control the development of debris flow and reduce the size and frequency of debris flow. Its main function is: (1) stop the flow of sediment and reduce the flow velocity, severe and scale of debris flow; (2) raise the erosion baseline of the local channel section and protect the solid slope of the bed; (3) slow down the longitudinal slope of the silt bed, decrease the erosion and impact of debris flow, reduce the erosion of gully bed, and inhibit the development of debris flow; (4) the increasing of downstream scour of the dam is beneficial to the sediment transport, which plays a regulating role in sediment deposition and channel evolution.

The dam may be set up to form a cascade dam, also known as the valley cluster. The dam should be used in combination with other measures. The dam belongs to semi-permanent engineering, and other renovation works can be used once the solid material is filled with overflow dam.

Grilled dam is the general term for the dam with horizontal or vertical grille, plane or three-dimensional grid and the overall framework structure, which mainly is suitable for the prevention and control of loose debris flow, does not apply to prevention and treatment of viscous debris flow. The grilled dam can be constructed with steel rails, steel bars and cables, and the spacing of grille shall be determined according to the situation of debris flow.

7.6 Substructure in karst area

7.6.1 The soluble rock formations such as limestone, generate special geomorphological morphology and hydrogeological phenomena under the long-term dissolution and denudation of the water. These kind of geomorphological morphology and hydrogeological phenomena is collectively referred to as karst. The damage that caused by karst to substructure is that: the substructure subsidence and destruction caused by the collapse of karst roof. The failure of karst ground collapse damages the stability of substructure; repeated springs and geysers soak the substructure, causing substructure subsidence, shooting collapse or slurping; the sudden burst of water burst the substructure. Firstly, it is necessary to find out the development law and distribution law of karst in geological conditions. Then, determine the layout and location of the route carefully. In general, it is necessary to take the measure of avoiding of bypassing the karst area which is of much harm, large scale, and difficult to find out. For small and median karst area where the harm is lighter, substructure should be constructed in the areas where the scale of karst is relatively small and easy for treatment.

The substructure design in karst area is mainly for the prevention and treatment of karst and karst water that affect the stability of substructure. The practice proves that, if not handled or handled improperly, it will not only produce various substructure diseases, but also affect the safety of driving. And it will lead to limited use of water resources, which affects local production and the normal order of life.

7.6.2 Factors in two aspects should be analyzed when evaluate the stability of the roof of karst cave: the first is the internal factors, including the thickness of the roof, span and morphology, rock properties, rock occurrence, the condition of joints and fissures, and rock physical and mechanical indexes, etc.; the second is the external factors, including the condition of the load and the mechanical damage of the water flow in the hole. There are many evaluation methods for roof safety thickness, and the original *Specifications* (JTG D30—2004) adopts the thickness ratio method which is based on the railway scientific research results and the practice experience in the northern Guangdong section of the Beijing-Zhuhai motorway. The engineering practice shows that the evaluation method and control standard are reasonable. So they are not revised this time.

7.6.3 The karst caves nearby the route requires a certain distance from the foot of the substructure, will not endanger the substructure at the time of the collapse of the cave. The distance is called the safe distance of the karst caves and the route. The calculation equation listed in the content is calculated according to the diffusion angle at the time of collapse. As there are many factors, it is recommended to investigate the examples of the existing projects in practice.

7.6.4 The karst water is one of the main factors that endanger the safety of substructure, so priority needs to be given to directing the karst water. For karst rise spring, it is not suitable for blocking. The engineering practice shows that it tends to cause the substructure slurry, slope collapse and other diseases if karst rise spring is blocked. In karst location, surface water and groundwater have a strong erosiveness, which is the main factor to cause dissolution and destruction of rock formations. Therefore, the substructure design should be careful to adjust the surface flow, dredging the water and groundwater. It is generally used to construct drainage channels, seepage channels, culverts and drainage tunnels to direct the water to prevent surface water and groundwater from harming the substructure.

7.6.5 When substructure is in the enclosed karst depression, the original surface water system is often cut off, and some embankments are directly covered with the water holes, which can easily cause substructure diseases. Therefore, it is very important to discharge the surface water in the dissolute depressions. It is necessary to have a good drainage design to ensure the smooth surface water based on the investigation of the water situation. For the embankment which is inevitable to be soaked in the water of the rainy season, the soaked parts need to be given priority to. Gravel and hard rock fragments with high water stability need to be chosen as substructure filling.

7.6.6 Karst cave, dissolution and fracture zone, and covering layer of soil holes, which endanger the safety and stability of substructure, should be treated according to the specific conditions. Backfill, avoiding, reinforcement, and other measures need to be taken in accordance with local conditions.

The backfill treatment is suitable for the dry karst cave which is shallow and has no discharge requirement. Stone and gravel are usually chosen as fillings. Cave fillings are usually loose, weak, cave surface corrosion part also relatively loose and broken. If use cave fillings as substructure base, their bearing capacity often can't meet the design requirements. Cave fillings need to be removed and replaced with fillings of high strength and water stability. When the top plate of the foundation is thin, the usual way is to blast the top plate of the hole; then remove the fillings and loose material; after that, backfill stone and gravel.

For small and deep caves, the steel reinforced concrete cover may be crossed according to its width. In the case of large - span karst caves, or the ones need to maintain drainage, a bridge or a culvert is generally adopted for passing. The bridge is suitable for crossing the river, running water holes or water holes. Culverts are suitable for crossing common karst springs.

When the diameter of the karst cave is large, the roof is complete and the construction condition is good, it is usually used to reinforce the wall or support column of the grouting stone or slabs. When the construction conditions in the hole are bad, it usually choose the bored piles to support. For dissolution fissure development zones, deep buried karst caves, and soil caves, it generally chooses grouting reinforcement for strengthening.

7.7 Substructure in the area of soft soil

7.7.1 Post-settlement settlement refers to the residual settlement of pavement designed years (asphalt pavement is 15 years, cement concrete pavement for 30 years).

7.7.2 Considering the influence of traffic load from the perspective of settlement calculation, it is necessary to consider the continuity of its action, because the occurrence of subsidence is not accomplished at a certain moment of load. The effect of the vehicle on the road is the intermittent continuous impact load, which has a certain difference with the static load at the time of action and the depth of the action. The measured data show that dynamic load additional dynamic stress in substructure decreased with depth, when the additional dynamic stress of a certain depth is 1/10 of gravity stress, this depth can be considered as the effective impact depth of traffic load. In calculating the settlement of soil layer below this depth, load dynamic stress can be neglected. The field test in the Lianyan highway conducted by Jiangsu transportation research institute co., LTD and Hehai university shows that the effective influence depth of a car (2 t) is about 1 m, and cart (20 t) is 2.5 m; Influence on the speed is small. *The Guidance of Soft Soil Foundation Treatment Technology* written by Japan Road Association (translated by Enjie Cai) thinks that low embankment under 2.0–2.5 m of the road surface often suffers from the harm of uneven settlement caused by traffic load. So 2.5 m can be regarded as the boundary between high embankment and low embankment when considering the effect of dynamic traffic load on the settlement: When the embankment height more than 2.5 m, it doesn't need to consider the effect of dynamic load on the settlement. When embankment height is less than or equal to 2.5 m, and the influence of dynamic load on the settlement needs to be considered.

The low embankment is not used much on the domestic highway, and the substructure average filling height of the substructure is generally between 3.0 and 3.8. Since April 2004 the ministry of communications issued *Several Opinions about implementing the strictest farmland protection system during highway construction*, the design concept of low embankment is taken seriously. There is no mature method to calculate the subsidence of soft soil based on the driving load on low embankment. In actual engineering, some measures are taken to control the settlement. For example, solidify the surface soft soil layer with lime or cement to generate artificial shell layer. Artificial shell layer not only has high resistance to deformation which improves the rigidity of low embankment, but also increases the ways for stress diffusion which makes the vertical stress passing to the soft soil layer significantly small; geosynthetic reinforced material can also be used in embankment to improve the rigidity and integrity of embankment and reduce the traffic load acting on the foundation. In addition, composite foundation and replacement are also effective methods to control dynamic load. Use layered summation method, compression test of *e-p* curve, compression modulus E_s or e-lg$p$ curve to calculate the consolidation settlement of soft soil foundation. It is the most commonly used method in engineering currently. There are much experience to use *e-p* curve and compression modulus E_s to calculate the accumulated settlement, and the corresponding settlement coefficient m_s value also has some experience, too. Based on the research results of the soft foundation test of Beijing-Tianjin-Tanggu express highway, the empirical equation of the calculation of the settlement coefficient m_s is given.

7.7.3 The effective consolidation stress method considers the actual situation of the construction of the soft foundation embankment, namely, the embankment is not to fill the design height in an instant, it is filled gradually according to certain construction rate. When it needs to build high embankment on the foundation which is of poor strength, this mode of calculation can be applied to check the increased safety coefficient after adopting the staged loading method to increase foundation consolidation strength. This may ensure that the stability meets the requirements in the process of embankment filling.

The improved total intensity method is developed based on the $\varphi = 0°$ method. It is based on the $\varphi = 0°$ method and the situ test data [Vane shear strength converted from the shear strength of static cone penetration test (single bridge probe) or static point resistance (double bridge probe) is adopted. Also, shear strength from vane shear test can be used directly.], borrow the idea to use the effective consolidation stress method to calculate the foundation strength increased with consolidation, use the strength growth factors to calculate the strength growth increment during the consolidation process. This method is combined with the static test to provide an efficient and reliable way for checking the stability of soft foundation embankment.

The simplified method of Bishop and Janbu are both relatively accurate method, and the Janbu method is often used in the stability calculation of non-circular sliding surfaces. Because the two calculation methods use effective shear strength index, the workload of the sampling test is relatively large. It is difficult to use this method during the entire process of the design. So this method may be selectively applied in test engineering or some critical parts of the embankment.

The calculation equations of the above four methods are as follows:

(1) When the effective consolidation stress method is adopted, the calculation equation of stability safety coefficient is:

$$F = \frac{\sum_{A}^{B}(c_{qi}L_i + W_{Ii}\mathrm{coa}\alpha_i\tan\varphi_{qi} + W_{IIi}\cos\alpha_i U_i\tan\varphi_{cqi}) + \sum_{B}^{C}(c_{qi}L_i + W_{IIi}\cos\alpha_i\tan\varphi_{qi})}{\sum_{A}^{B}(W_I + W_{II})_i\sin\alpha_i + \sum_{B}^{C}W_{IIi}\sin\alpha_i} \tag{7.1}$$

where:

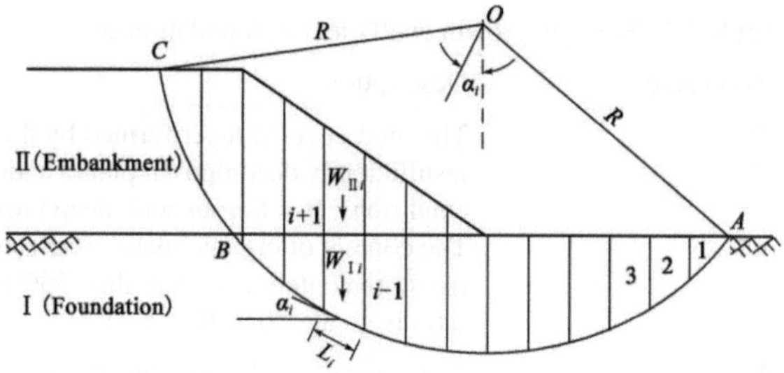

Fig. 7.1 Safety factor calculation schema

c_{qi}, φ_{qi}—the cohesion and internal friction angle of the quick shear test of foundation soil or embankment fillings;

φ_{cqi}—internal friction angle of foundation soil consolidation obtained by quick shear test;

U_i—average consolidation degree of foundation.

The rest of the symbols are shown in Fig. 7.1.

(2) When using improved total strength method, the calculation equation of stability safety coefficient is:

$$F = \frac{\sum_{A}^{B}(S_{ui} + W_{IIi}\text{coa}\alpha_i U_i m_i)L_i + \sum_{B}^{C}(c_{qi}L_i + W_{IIi}\cos\alpha_i\tan\varphi_{qi})}{\sum_{A}^{B}(W_I + W_{II})_i\sin\alpha_i + \sum_{B}^{C}W_{IIi}\sin\alpha_i} \quad (7.2)$$

where:

S_{ui}—vane shear strength or shear strength converted from static penetration resistance of static cone penetration test (single bridge probe) or static point resistance (double-bridge probe);

m_i—the strength growth coefficient of soil layer. It is referred to Table 7.1.

The rest of the symbols are the same to the above ones.

(3) When using simplified Bishop method, the calculation equation of stability safety coefficient is:

$$F = \frac{\sum_{A}^{B}\{c'_i b_i + [(W_I + W_{II})_i u_i b_i]\tan\varphi_i\}/m_{I\alpha i} + \sum_{B}^{C}(c_{qi}b_i + W_{IIi}\cos\alpha_i\tan\varphi_{qi})/m_{II\alpha i}}{\sum_{A}^{B}(W_I + W_{II})_i\sin\alpha_i + \sum_{B}^{C}W_{IIi}\sin\alpha_i} \quad (7.3)$$

$$m_{I\alpha i} = \cos\alpha_i + \tan\varphi'_i\ \sin\alpha_i/F \quad (7.4)$$

$$m_{II\alpha i} = \cos\alpha_i + \tan\varphi_{qi}\sin\alpha_i/F \quad (7.5)$$

Table 7.1 Strength growth coefficient of foundation soil

Soil name	Description	m_i
Peat	The mud-covered layer formed by the accumulation of insufficiently decomposed plants under wet and anoxic conditions. It is fibrous with deep brown to black color. The content of organic matter is greater than 60%, the moisture content is greater than 300%, and the porosity ratio is greater than 10	0.35
Peat soil	The majority of the remains of the plants is completely decomposed into a smelly, black, muddy soil. The content of organic matter is between 10 and 60% (subdivided into weak peat soil, medium peat soil and strong peat soil), with water content of no more than 300%, and the porosity ratio is greater than 3	0.20
Organic soil	The fine-grained soil composed of different decomposed vegetative plants in a multi-water environment, among which mineral particles are mixed. The organic matter content is between 3 and 10%, and silt and silty soil belong to this kind	0.25
Clayey soil	The plasticity index (76 g cone) is more than 17	0.30
Silty soil	The plasticity index (76 g cone) is greater than 10, but less than or equal to 17	0.25

where:

c'_i, φ'_i—The effective cohesion and effective internal friction angle of the foundation soil obtained by triaxial test;

b_i—The horizontal width of the strips, $b_i = L_i \cos\alpha_i$;

u_i—pore water pressure on sliding surface.

The rest of the symbols are the same to the above ones.

Because F is contained in $m_{\alpha i}$ which is in the right side of the stability safety coefficient formula, the calculation of safety coefficient needs to adopt iterative method.

(4) When using Janbu method, the calculation equation of stable safety coefficient is:

$$F = \frac{\sum_{A}^{B} \{ c'_i b_i + [(W_{\mathrm{I}} + W_{\mathrm{II}})_i u_i b_i + \Delta T_i] \tan\varphi'_i \} / m_{I\alpha i} / \cos\alpha_i + \sum_{B}^{C} (c_{\mathrm{q}i} b_i + W_{\mathrm{II}i} \cos\alpha_i \tan\varphi_{\mathrm{q}i} + \Delta T_i) / m_{\mathrm{II}\alpha i} / \cos\alpha_i}{\sum_{A}^{B} (W_{\mathrm{I}} + W_{\mathrm{II}} + \Delta T_i)_i \tan\alpha_i + \sum_{B}^{C} (W_{\mathrm{II}i} + \Delta T_i)_i \tan\alpha_i} \tag{7.6}$$

where:

ΔT_i—the increment of shear force on both sides of the soil bar. It may be calculated according to the assumption of the position of the force on both sides of the soil boundary.

The rest of the symbols are the same to the above ones.

Because F is contained in $m_{\alpha i}$ which is in the right side of the stability safety coefficient formula, the calculation of safety coefficient needs to adopt iterative method.

7.7.6 Drainage consolidation method is commonly used in soft soil foundation treatment, including sand cushion preloading, plastic drainage plate or bagged sand drain preloading and vacuum preloading and combined vacuum and surcharge preloading. Drainage consolidation method is often used together with lightweight embankment and reinforced embankment, loading berm, etc.

From the Chengjiang observation test engineering results of the soft soil foundation Beijing-Tianjin-Tanggu express highway, the central settlement of the embankment which is set with sand bags or plastic drainage plate back to normal after 4–5 months. While the settlement of the road shoulder returns to normal after 6 months. So restricting preloading period to avoid from too brief time is necessary. *The Design Specification of High-grade Highway* of Japan makes it clear that the pre-pressure construction needs to be put in place for more than 6 months in principle. From the perspective of construction period and the economy, six months is the proper time. To ensure the effect of vacuum preloading, the permeability coefficient of the reinforced soft soil layer is generally required to be 10^{-5}-10^{-6} cm/s; when the reinforced area is permeable to air and water (horizontal sand layer exists), permeability coefficient cannot satisfy the required conditions, clay mixing pile wall can be used to enclosed the consolidated area, also can increase the number of jet pump or increase power. For example, the container yard in deep-water port area of Shantou Port adopts vacuum combined pile load preloading to treat the 57,248 m^2 soft soil foundation from 1994 to 1995. Because there are 4–12.5 m fine sand and coarse sand layer on the surface, 1.2 m clay sealing wall was built for sealing. Vacuum pressure reaches more than 80 kPa after 2-week suction. The maximum depth of the plastic drainage plate is 25 m. The soft soil layer of the foundation of the Zhangwan interchange ramp in Funing expressway has a horizontal sand layer and the subsoil in the subsurface is mixed with pebbles. After water injecting test, the permeability coefficient of pebble mild clay was 2.5×10^{-4} cm/s. But by increasing the number of jet pump, the vacuum degree was kept at 70 kPa and the foundation reinforcement effect was good. Although the vacuum degree is large or small, it can be used to strengthen the foundation. But the vacuum is less economical when the degree of vacuum is low. In the engineering, the vacuum degree under the airtight membrane is generally required to be 70–90 kPa.

The calculation method of foundation settlement under the load of the combined vacuum and surcharge preloading is the same to the calculation method of foundation settlement under the load of the preloading. The difference is that the load of preloading is replaced with vacuum degree plus the load of embankment fillings. Because vacuum preloading produces a negative pressure with the horizontal direction to negative pressure source, the surrounding soil moving toward preloading area, produce isotropic consolidation. It won't produce shear deformation. Shrinkage deformation will occur, so the combined vacuum and surcharge preloading can offset the lateral soil extrusion deformation produced by preloading, good for the stability of the foundation, generally will not occur instability and failure of foundation.

7.7.7 Strengthening soft soil foundation with granulated pile is used for replacement, drainage consolidation and stress concentration.

The length of aggregate pile within the foundation belongs to the composite foundation and the basic assumption of composite foundation theory is the deformation coordination of pile and soil, it is generally not considered about negative frictional resistance of pile and pile group effect in the design.

Aggregate pile bearing capacity is not only related to the nature of the pile body material and the density of the pile body, but also related to the side of the limit of ability, when the strength of reinforcement of soft soil is very low, it is difficult to aggregate pile. For the applicable conditions of the vibro-re-placement gravel pile, *Technical Code for Ground Treatment of Buildings* (JGJ 79) stipulates that the undrained shear strength of the reinforced soil is not less than 20 kPa. The domestic research results show that the foundation soil pile can still use vibration displacement method for undrained shear strength is slightly lower than 20 kPa. In recent years, adopting measures such as large diameter coarse particle size gets more successful project in Guangdong and Fujian. It is feasible to use vibro-re-placement process in foundation soil with the undrained shear strength is over than 15 kPa.

The construction of immersed tube method is hard to use in the low strength of soft soil foundation because of the large disturbance and no breast wall. When the vane shear strength is less than 20 kPa it should not be used according to engineering experience.

The construction of granule pile will cause disturbance to the land. It is generally considered that the soil strength may be reduced by 10% to 40% and the strength can be restored within 20–30 d during construction of the vibration and shock displacement method. The strength of the silty soil can be recovered over 30 d when using the submerged tube method in the construction of the soil, and the influence of this factor should be paid attention to when the foundation strength is tested.

As for the design parameters, the Japanese data suggest that the angle of internal friction is 25° for sand pile, 35° for gravel pile, and pile soil stress ratio is 3. Some domestic empirical formulas for calculating the bearing capacity of gravel pile mostly use 35°-40° for the internal friction angle of gravel pile. The stress ratio of soil between granule piles is different from that of soil and depth and the size and duration of load. Most of the data believe that the n value is appropriate between 2 and 5.

7.7.8 The shear strength of the age of 90 d for the reinforced soil pile is considered as standard strength, which is in line with the construction of highway embankment at present. After the construction of the pile over the age the 90 d, the embankment may be filled to around 3 m. The load is not large, and the stability calculation is still safe.

Choosing 90 d's strength as standard strength brings difficulties in the laboratory test and field test. One solution is that based on short age test and the testing data (7 d, 28 d) to speculate the strength of the 90 d according to the law of strength increasing (Table 7.2 empirical formula can be referenced). The second method is that adopting high temperature and rapid health so as to achieve the standard health of 90 d in a very short period of time. According to research results of China first highway survey design and research institute (FHCC), high temperature curing 30 h is equivalent to the strength of the standard curing 28 d, high temperature curing 96 h is equivalent to the strength of the standard curing 90 d.

The compression modulus test for cement reinforcement soil pile is more complex than the unconfined compressive strength test. It may calculate according to the experience formula of the unconfined compressive strength, however empirical formula is quite different, such as $E_p = (100 - 120)q_u$ (for cube specimens), $E_p = (25 - 50)q_u$ etc. In addition, the unconfined compressive strength for cube specimen is different from measured by the cylinder, which is less intense than the former. According to the research of the first highway survey and design research institute of China, indoor cement reinforced soil mix ratio test should be suitable for the round test mode ($\varphi \times h = 50$ mm $\times$ 100 mm), which is consistent with the field drilling core. The relationship between the compressive strength and the deformation modulus of the unconfined compressive strength is as follows: $E_p = 83.4q_u$ (for cement reinforcement soil, approximate compression modulus equals deformation modulus); the relationship between the unconfined compressive strength of round test mode ($\varphi \times h = 50$ mm $\times$ 100 mm) and the strength of square test mode (70.7 mm $\times$ 70.7 mm $\times$ 70.7 mm): the strength of the round test module = 0.87 * the strength of the square test module.

Although the reinforced soil pile uses lime as the curing agent, but as a result of using lime powder (oar) for curing materials mixing pile in expressway soft foundation processing is rarely adopted, the contents of the provisions cannot be used for lime mixing pile. Special tests are needed when reinforced soil pile is used.

7.7.9 Steps for proportioning design of Cement Fly-ash Gravel Pile (CFG pile):

(1) Determine water consumption W

The water consumption is determined by the specific value of slump, and generally starts from the empirical water consumption; make water consumption per cubic meter be W.

(2) Determine the amount of cement C

According to the cement strength grade R_c^b, the 28 d strength of the mixture is f_{cu}, and cement consumption per cubic meter is calculated according to the following equation:

$$f_{cu} = 0.366R_c^b\left(\frac{C}{W} - 0.071\right) \quad (7.7)$$

where:

f_{cu}—28 d strength of the mixed material (MPa), tested from a regular cube with the length of 150 mm;

R_c^b—the strength level of cement (MPa);

C—cement consumption per cubic meter (kg);

W—water consumption per cubic meter (kg).

Table 7.2 Comparison of strength and age of cement mixing soil in different areas

Representative area and source of information	Relational formula	Suppose that strength of 28 d is 1.0 MPa calculate according to the relational formula		Note
		Strength of 7 d (MPa)	Strength of 90 d (MPa)	
China first highway investigation and design institute (test samples taken from Tianjin, Fujian, Lianyungang and Nantong)	$q_{u28} = 2.37q_{u7} - 0.19$ ($r = 0.87$, $n = 12$) $q_{u90} = 1.14q_{u28} - 0.85$ ($r = 0.79$, $n = 15$)	0.50	1.99	q_{u7}, q_{u28}, q_{u90} represent unconfined shear strength of 7 d, 28 d, and 90 d r, S, n represent correlation coefficient, standard deviation and statistic group respectively
Technical Code for Strengthening Soft Soil Layer by Powder Spraying Method (TB 10,113—96)	$q_{u28} = 1.49q_{u7}$; $q_{u90} = 1.97q_{u7}$; $q_{u90} = 1.33q_{u28}$	0.67	1.33	
Tianjin region, Tianjin harbor engineering research institute, *Study on the characteristics of cement identification soil engineering* (research report)	Silt: $q_{u7} = 0.364q_{u90}$ $q_{u28} = 0.652q_{u90}$ Muddy Soil: $q_{u7} = 0.262q_{u90}$ $q_{u28} = 0.485q_{u90}$	0.56 0.54	2.54 2.06	
Shanghai area, *Foundation treatment* (Shulin Ye)	$q_{u7} = 0.56q_{u28}$ ($r = 0.98$, $S = 0.059$, $n = 15$) $q_{u90} = 1.63q_{u28}$ ($r = 0.98$, $S = 0.143$, $n = 9$)	0.56	1.63	

(3) Determine the consumption of fly ash F

The consumption of fly ash per cubic meter is calculated according to Eq. (7.8):

$$\frac{W}{C} = 0.187 + 0.791\frac{F}{C} \tag{7.8}$$

(4) Consumption of stone chips G_1 and consumption of gravel G_2

The stone chip rate λ is needed to calculate consumption of stone chips per cubic meter G_1 and consumption of gravel per cubic meter G_2:

$$\lambda = \frac{G_1}{G_1 + G_2} \tag{7.9}$$

Table 7.3 Effective reinforcement depth of the dynamic compaction method (m)

Tamping energy per tamping (kN · m)	Gravel soil, sand and other coarse grain soil	Ramming silt, clay, collapsible loess and other fine grain soil
1,000	5.0–6.0	4.0–5.0
2,000	6.0–7.0	5.0–6.0
3,000	7.0–8.0	6.0–7.0
4,000	8.0–9.0	7.0–8.0
5,000	9.0–9.5	8.0–8.5
6,000	9.5–10.0	8.5–9.0

Note The effective reinforcement depth of the dynamic compaction method shall be calculated from the initial ramming surface

According to the experimental results, the value of λ is 0.25–0.33, which is the reasonable stone chip rate.

The density of mixed materials is known (generally 2.2–2.3 t/m^3), and $G_1 + G_2$ can be calculated through the already obtained W, C and F, and then the G_1 and G_2 can be obtained respectively through Eq. (7.9).

Adjust the water consumption according to the slump and the above steps until the requirements are met.

7.7.10 Effective reinforcement depth managed by dynamic compaction can be calculated using the Menard empirical formula, but the estimated results are indeed different from the actual situation. The correction coefficient needs to be used during the design process. The correction coefficient is related to the soil condition, underground water level, the size of the ramming energy and the base area of the ramming hammer, which is generally 0.34–0.80. The *Technical Specification for Building Foundation Treatment* (JGJ 79) can be referenced by the foundation soil class and the effective strengthening depth listed by the tamping hammer, which is shown in Table 7.3.

The best tamping energy can be determined by observing the pore water pressure or ramming during the test compaction process to determine the optimum tamping energy (ramming ramming). Due to the observation of ramming is simple and easy, so the project used in the number of ramming and ramming curve to determine the best tamping energy.

7.7.11 Highway soft soil foundation treatment adopts the methods of rigid pile including prestressed concrete thin-wall pipe pile (PTC), prestressed concrete pipe pile (PHC), precast concrete piles and bored piles, cast-in-situ thin-wall tubular pile, and so on, the most commonly used is the prestressed concrete thin-wall pipe pile. The pipe pile is prefabricated pile from the factory. The outer diameter of the pile is usually 300–500 mm, the wall thickness is 60–100 mm, the pile length is standardized and customized. They can be connected by welding during the site construction process. The cast-in-situ concrete thin-wall pile is to infiltrate the double-layer casing into the soft soil foundation, and the concrete is poured into the double casing to form a large diameter cylindrical pile. Its diameter is generally 0.8–1.5 m, and the wall thickness is

Table 7.4 Physical and mechanical indexes of red clay

Water content (%)	Void ratio	Liquid limit (%)	Plastic limit (%)	Saturation (%)	Water content ratio	Coefficient of compressibility (MPa^{-1})	Permeability coefficient (cm/s)	Free swelling ratio (%)
20–75	0.7–2.1	40–110	20–60	80–100	0.5–0.75	0.1–0.4	$i \times 10^{-8}$	25–69

120–200 mm. Rigid pile cushion layer is made of synthetic materials and sandstone with different composition in the form of different pavement, the main types are: Geogrid reinforced soil cushion, geocell cushion, high strength composite geotextile reinforced cushion, high strength geotextile, long pipe bag, reinforced cushion, and so on. These types shall be selected according to the load size and requirements of the design project and the conditions of the soil layer of concrete foundation.

7.7.15 The settlement rate at any moment and the settlement after the work may be determined according to the settlement observation curve. Regarding the settlement rate as settlement stability control standard is the widespread practice of the highway in soft soil area in China. The project practices prove to be effective. For example, Beijing-Tianjin-Tanggu motorway removed the preloading soil when the settlement rate reaches 8 mm/month to start the construction of pavement structure; the settlement rate of Lianxu motorway and Funing motorway is 5 mm/month. Shanghai-nanjing motorway regards the top surface of road trough as preloading calculate height. When the settlement rate reaches 5 mm/month, the preloading soil is unloaded to construct subbase and roadbase. The preloading is continued after the roadbase is constructed. The top course of the road is constructed when the settlement rate reaches 3 mm/month.

Overload preloading has been widely used as a loading method. Due to over consolidation characteristics of soft soil, there would be much different in settlement rate of surcharge preloading foundation before and after unloading (Generally differ by 3 to 5 times, the greater the overload ratio is, the greater the difference will be), the settlement rate will decrease after unloading. So the 5 mm/month means the settlement control standard under the same loading conditions.

7.8 Substructure in red clay and high liquid limit soil area

7.8.1 Red clay refers to highly plastic clay, the color of which is brownish red and brown-and-buff, formed from the rock exposed by carbonate rocks by lateritization. Its liquid limit is generally greater than 50%. After re-transport, the clay still remains the basic characteristics of red clay, and the liquid limit of which is greater than 45%, is called secondary red clay.

Red clay is a kind of regional special soil including slope deposit and residual soils type. Its physical and mechanical properties refer to Table 7.4.

Red clay and high liquid limit soil have the characteristics of high liquid limit, high plasticity, high water content, high void ratio, high strength and low compressibility. The difficulty of compaction, shrinkage cracking and slope stability are the main problems of engineering. The swelling potential of red clay is low, the expansion rate without load is less than 20%, and the expansion pressure is generally less than 30 kPa, therefore the swelling is weak; the linear shrinkage rate of red clay is 1–10%, the volume shrinkage ratio is 5–28%, and contraction coefficient is 0.1–0.8, which was weak to moderate shrinkage.

The soil, the liquid limit ratio of which is greater than 50%, is called high liquid limit soil, its causes are complex, mainly related to the parent rock properties. The engineering properties of high liquid limit soil are different from red clay, with higher bound water content and more obvious separation phenomenon of strength and compactness compared with red clay.

7.8.2 Under the influence of climate, environment (rainfall, humidity and temperature), surface water and ground water, the strength of cement soil is directly destroyed and resulted in deformation and failure of substructure when the substructure is filled directly with red clay and high liquid limit soil. 0.5 MPa^{-1} is filled in the substructure, during the operation of the road, the substructure will creep and the uneven deformation of the substructure will be increased. Therefore, the *Specifications* stipulates that red clay and high liquid limit soil cannot fill substructure directly and limits the use of red clay with compression coefficient greater than 0.5 MPa^{-1}.

7.8.3 In order to prevent the disease of red clay and high liquid limit soil, when red clay and high liquid limit soil are used for substructure fillings, some engineering measures need to be taken to ensure the stability of long-term performance. Common engineering measures include physical measures and chemical treatment measures. The measures include setting up drainage isolation cushion, wrapping and sealing layer, and adding gravel, fly ash, crushed stone and so on. The chemical treatment measures mainly are adding lime, cement and other inorganic combination materials to improve the treatment.

For lime treatment of high liquid limit soil and red clay, the'two times mixing ash' method is usually adopted. For the first time, the soil is mixed with a mixture of 1/2–2/3 design of lime, which is mixed with ash, and then the soil is tossed to the ground and 1–3 d is piled in the soil field. Then, the soil material with the first lime treatment will be transported to the substructure, and then the second lime treatment will be added, the dosage of ash is the remaining 1/3 design dose.

The high liquid limit soil has a high content of fine particles, and its internal colloidal substance ($Fe_2O_3 \cdot nH_2O$, $SiO_2 \cdot nH_2O$ and so on) contains the binding water. The bound water is a part of the material particles, which is different from the free water of the ordinary soil. After the drying of the high liquid limit soil, the binding force and molecular structure between the bound water and the particles are destroyed, and the gelling effect of the liquid limit cannot be recovered after filtration. The test results of wet and dry process have larger gap, under the same conditions, the wet method of the specimen is higher than that of dry, undisturbed soil strength is higher than that of disturbed soil strength, compaction test of the maximum dry density is less than that of dry, wet than dry, wet optimum moisture content is greater than the dry method. In situ, natural moisture content of soil packing with high liquid limit is generally larger. Therefore, the packing must be compacted by layer after decreasing moisture content by drying. Compaction test and CBR test method adopt wet parts, which is more in line with the time construction process.

7.8.5 The red clay with high liquid limit soil slope excavation, in the alternate effect of hot and humid climatic conditions, soil would shrink and craze, so the fissure in red clay is developed. Red clay with strong contractility, in the topography, with a sunny exposure and less vegetation, the fracture density is large, extending are approximately 3–4 m deep, and more than 10 m in individual areas is more than 10 m. The fracture destroys the integrity of the soil, reduces the strength of the soil, increases the permeability of the soil, and constitutes a negative factor of soil stability. Rainfall results in the infiltration of rainwater along the cracks, forming a weak structural plane of the soil. Even if the slope ratio is less than 1:2, there may be collapse and sliding failure, and the shear exit of the landslide is more than the top of the substructure. The destructive mode of the red clay excavation has obvious difference with the arc sliding of ordinary soil.

According to the characteristics of the slope damage of red clay and high liquid limit soil excavation, the specification states that stability checking should be carried out when the height of the excavation is more than 10 m. In slope stability analysis and calculation, it needs to fully consider the red clay slope crack development and the complex flooding adverse impacts on the slope stability, shear strength parameters need to use strength indexes of water test and repeated consolidated drained direct shear test and others. As it is available, the first wet dry cycling test needs to be constructed on soil samples, then the saturated shear test needs to be taken.

The instability of red clay slope excavation is mainly caused by crack ooze water. Therefore, the comprehensive design of red clay slope protection and drainage needs to be strengthened in the design and assume the change of the slope soil moisture and arise and development of its cracks. The engineering practice shows that the combination of supporting seepage groove slope and arch slope support, is the effective measure to ensure the slope stability of red clay.

7.9 Substructure in expansive soil area

7.9.1 Expansive soil is a kind of high plastic clay, which is hydrophilic minerals, and have clear water absorption expansion and water shrinkage characteristics. The expansion characteristics of expansive soil are mainly influenced by montmorillonite clay minerals with crystalline structure. Under the influence of atmosphere, humidity change causes expansion and shrinkage of expansive soil, soil cracking, rainfall infiltration, intensity reduction, and large expansion pressure, resulting in deformation and destruction of slope. In the shallow layer, the maximum effect of atmosphere on expansive soil is called expansive soil activity zone depth.

According to the characteristics of the expansive soil, expansive soil area substructure design needs to give priority to waterproof, moisture, prevent weathering, and take effective measures to reduce the influence of humidity changes of expansive soil substructure, guarantee the substructure to meet the requirements of the deformation and strength. In the construction of expansive soil substructure, it is also should be continuous construction, and timely closed the road bed and slope surface.

7.9.2 The criteria for determining the expansive soil are not uniform around the world. According to experience of many engineering practice for many years and engineering geological characteristics, when free expansion rate is more than 40%

and liquid limit of the clayey soil is greater than 40%, it may be defined as expansive soil preliminary, but this is not the only factor. The final decision factor is swell-shrink total rate and the expansion of the cyclic deformation characteristics.

The Expansive Soil Area Architecture Specifications (GB 50,112—2013) adopts the free expansion rate to classify the expansion potential of the expansive soil. Free expansion rate δ_{ef} test, which uses grinding fine sieving dry soil, and dry it in 105–110 °C to a constant weight. After cooling in the dryer, the operator uses standard measuring cup take 10 cm^3 soil samples to test the free expansion rate. There has been controversy over the reliability of the index and the extent to which it reflects the nature of expansive soil. Some people think that the test method causes the intergranular joint to be lost, and the expansion is well played, and the results do not represent the actual expansion potential of the soil.

The properties of the expansive soil are determined by the amount of the inner montmorillonite and its family. The free expansion rate discrimination method produces the misjudgment and omission of expansive soil. The railway department uses montmorillonite content and cation exchange capacity as the identification index, which can judge the high accuracy, but the test is difficult.

There is a linear correlation between the moisture content of the standard and the ratio of surface area, cation exchange volume and montmorillonite. The standard moisture content reflects the most basic essence of expansive soil. Standard moisture absorption water content test method has been incorporated into *The Highway Soil Test Specifications* (JTG E40—2007). The discriminant standard has been widely used in highway industry and construction industry in Hubei, Hunan, Anhui, Guangxi, Yunnan, Henan, Hebei, and is widely applied in The water conservancy project, verifying the accuracy and reliability of the standard.

7.9.3, 7.9.4 Expansive soil foundation deformation calculation is based on expansive soil foundation in the absence of construction creation, considering the extreme expansion of land base table moisture content change, and the change of moisture content along the depth distribution, which represents the inflationary expansion soil strength and the degree of response to climate.

Expansive soil foundation deformation prediction is the key to determine the activity of expansive soil depth under the influence of atmosphere. The depth of the expansive soil zone can be measured by the curve, which is changed with depth, such as ground temperature, soil moisture content under season change, or use the static cone penetration test methods such as penetration resistance changing with depth curve proven activity of expansive soil depth. The expansive soil activity area in parts of China is shown in Table 7.5.

According to the superstructure type and expansive soil base deformation, the soil treatment of expansive soil can be used to remove expansive soil, replace non-expansive soil or lime modified treatment. The latest research shows that the depth of the expansion soil is not less than 2.0 m in non-expansive soil or lime modified, and the water content, strength and density of the lower swelling soil are very small and negligible. When the water permeability material is used, the composite geomembrane need to be laid as the impermeable layer.

Table 7.5 Depth of active area in some typical expansive soil areas in China

Region	The critical activity zone depth of expansive soil under various determination signs (m)				The depth of the active zone (m)
	Temperature indicator	Geothermal Indicator	Depth indicator	Ground fissure indicator	
Jijie, Yunnan	3.0	–	–	–	3.0–4.0
River pool, Yunnan	5.0	–	–	–	3.0–5.0
Chengdu, Sichuan	1.5	1.8	–	–	1.5
Nanning, Guangxi	2.0–3.0	–	3.0	2.0–2.5	2.5–3
Ningming, Guangxi	–	–	3.5	2.5–3.5	3.0
Ankang, Shaanxi	3.0	–	–	2.0–3.0	3.0
Jingmen, Hubei	1.5–2.0	2.0	1.5	1.2–1.5	1.5–2.0
Yunxian, Hubei	2.0	2.0	–	< 2.0	2.0
Yichang, Hubei	–	2.1	–	–	2.1
Nanyang, Henan	–	3.2	–	–	3.2
Pingdingshan, Henan	2.5	2.1	3.0	–	2.5
Hefei, Anhui	2.0	2.0	–	–	2.0
Handan, Henan	2.0	–	–	–	2.0

7.9.5, 7.9.6 Expansive soil substructure filler, after compaction of the engineering properties of expansive soil and natural undisturbed expansive soil, has the very big difference, mainly the expansion of the compacted soil is larger than the original expansive soil swelling 5–8 times, some even reached 20 or 30 times greater. The greater the density of the filling soil, the lower the water content, the larger the swelling and swelling force after the soil is immersed in water. Under the same compaction moisture content, the higher the density, the larger the expansion and expansion force.

The expansive soil is used as a substructure packing, and different treatment measures need to be taken according to the scale of expansion. When it is weak expansive soil, physical treatment measures after filling substructure, which is the selection of expansive soils or inorganic binder, and set cushion at the edge of the embankment, or on both sides of the bag closed, and at the top of the capping layer physical measures, such as climate control and groundwater effect on expansion of dirt road dike humidity change, to guarantee the stability of substructure; for medium expansion soil, it is necessary to use inorganic binding material to fill the substructure. Strong expansion soil cannot be used for substructure packing. The *Specifications* experiences in expansive soil area highway construction. According to the characteristics of expansion road dike, the *Specifications* proposes expansion road embankment structure, typical expansive soil treatment measures and technical requirement of waterproof and drainage. In the design, it is necessary to select reasonable engineering measures according to the scale of the expansion soil, select reasonable engineering measures, prevent and control the damage of the dike, and guarantee the stability of the substructure.

Lime mixing is the most effective method for modification of expansive soil. The experimental study shows that, in general, the lime dose should be controlled at 4% to 10%. The optimum ratio of lime is treated as non-expansive soil under the low limit index of weak expansive soil.

7.9.7 Expansive soil slope design is a relatively complicated engineering geological problem. According to the current survey results, the slope rate of 1:2 -1:3 is generally adopted, but it also appears unstable, especially when there is a weak interlayer, and the slope adopts 1:5–1:8, which is not necessarily stable. The slope rate of slope is not the only factor, and it is not proper to solve the slope stability of the expansive soil.

The destructive form of slope of expansive soil is diverse, but from the depth of damage, it may be summarized as shallow layer and deep damage. Shallow damage refers to the deformation caused by the atmospheric influence layer, which is mainly affected by climate change, weathering degree, fracture development degree, etc., and is the main form of the failure of the slope of expansive soil. The deformation of the slope beyond the thickness of this layer is the deep damage, which is mainly caused by the adverse structure of the slope. When designing, the specific geological conditions of the slope need to be treated separately.

For a long time, expansive soil substructure slope excavation always uses rigid protective measures, such as retaining wall, sheet-pile wall, which cannot ease the expansive soil swell-shrink deformation of expansive force, especially in the strong force of the expansion of the expansive soil, and often makes the deformation of the retaining structure damage. In addition, the retaining structure of retaining wall and pile wall is incongruous with the surrounding environment.

In recent years, Western transportation construction project 'expansive soil area highway complete construction technology research', aiming at expansive soil cut slope failure mechanism, study and put forward the prevention and control of expansive soil slope deformation and failure of flexible supporting technology, package edge, and in Nanning, Guangxi Nanyou highway, which has been successfully applied to Baise highway can both solve the problem of expansive soil cut slope stability, and coordinate the road with the surrounding environment. The *Specifications* includes non-expansive clay coating technology or flexible support structure of expansive soil excavation. When designing, designers should according to the specific situation, especially negative structure between the expansive soil layer and the underlying rock layer, and apply the *Specifications* flexibly.

The improvement of the substructure drainage facilities directly affects the long-term performance and stability of the expansive soil substructure, such as waterproof and moisturizing, which can reduce the harmful effects of the expansion and drying shrinkage of the soil. In the design, according to the engineering properties of expansive soil, the designers should set up perfect waterproof and drainage system, to prevent the infiltration of surface water and groundwater to substructure ontology or cutting slope, and the maintenance of natural moisture content of soil condition is relatively stable.

7.10 Substructure in loess region

7.10.1 Loess is a yellowish red clay primarily composed of silt, featuring high porosity, low natural moisture content and calcium-containing. The total area of China's loess region accounts for over 6% of the entire land area. China's loess region mainly distributed in the arid and semi-arid inland areas, locating at 34°N–41°N. Within this range, the loess of loess plateau in the following areas is most typical: the central Shaanxi Plain, northern Shaanxi Province, Ningxia Province, western Henan Province, eastern Gansu Province and central Gansu Province. These regions are all located in the middle reaches of the Yellow River, which from the north of Qinling mountains (32°N–34° N) to the south of the Great Wall (46°N) in latitude, and from the west of Taihang Mountain (110°E–110°E) to the east of Riyue Mountain (94°10'E–103°04'E). The loess of these regions has characteristics that continuous distribution and thick soil layer, primarily consisted of eolian loess.

The collapsibility of loess reflects the collapse deformation caused by external load or dead load after being soaked. Collapsibility declines with the increase of depth, water content and dry unit weight, and the decrease of void ratio. It tends to disappear when the depth is over 10 m, dry unit weight is beyond 15 kN/m^3, or void ratio is lower than 0.8. Old loess has no collapsibility while new loess has collapsibility or strong collapsibility. Generally speaking, the loess in drift bed, pluvial loess or newly accumulated loess all has collapsibility. The collapsibility of the loess in slope cutting and eolian deposit loess is stronger than the one of alluvion loess and pluvial loess. By compression test, the collapsibility of loess may be defined as non-collapse loess or collapse loess, self-weight collapsible loess or non-self-weight collapsible loess. Meanwhile, the degree of collapsibility can also be clarified.

The drainage system of the substructure in loess region, as well as the design of protection engineering needs to be in line with the aims of erosion protection, anti-seepage, soil and water conservation and environment protection. Early delivery and far bring are the approach to these terms while dealing with outfall and inlet is the key of issue. Otherwise, several problems would be caused, including slide of slope, Broken gullies and damp sink of foundation.

When building roads in loess region, loess will be booth the subgrade of road and the filler of substructure (Old loess has a large amount of clay, with low permeability. It will be softened after touching water and the strength will be lowered rapidly. Therefore, the shape of substructure is easy to be changed, and the road shoulder with slope is possible to collapse. Old loess, thus, is not suitable to be the filler of substructure.). Due to the uniqueness of loess engineering, the knowledge on loess needs to be improved when conducting road engineering, so as to ensure that effective measures can be taken to ensure the stability of roadbed, substructure, and the cutting slope.

7.10.2 Though the high road embankment in loess region are common, the entire deformation of side slope seldom happened. This reflects that the current section form and slope gradient of high road embankment is suitable. Only if compaction quality can meet design requirement, the entire stability of side slope can be ensured.

When the height of fill slope is higher than 30 m, substructure has the possibility to have relatively serious deformation. Such deformation will bring more difficulties on engineering and maintenance. Therefore, in terms of design, we need to make a decision after a comprehensive comparison with bridge schemes from the perspectives of engineering cost, engineering difficulty, maintenance and settlement. Based on the calculation and analysis on slope stability, we need to define the ratio and shape of slope, integrating topography, stratigraphy and hydrology, in terms of the location requiring embankment basis.

To ensure the substructure compactness of substructure and reduce the settlement after engineering, in terms of high fill loess substructure, project cases in china will give supplement pressure when filling height is over 10 m. Such a supplement approach can be conducted by impact rolling, tension-weight-consolidation and dynamic compaction.

7.10.3 In terms of formal design of excavation slope, the stability, durability of side slope and the efficiency of Excavated section need to be taken into account, as well as the requirements on engineering and maintenance. The type of side slope need to be decided after comprehensive consideration, including the time causes of loess, located geographic units, tectonic joints, the height of slope, the conditions of surface water and groundwater, and the shape of natural slope.

According to engineering experience and study results, typical form of loess cutting slope and its slope gradient have been presented in Tables 7.15 and 7.16, which is suitable for homogeneous soil and the road section without poor engineering geologic phenomena. In the process of designing, decision needs to be made after the analysis on the location and specific condition of slope. In terms of deep cuttings with the depth over 30 m, it tends to be economically efficient for expressways, arterial highway and tunnels to pass through cuttings.

Setting a big platform at the middle of deep cutting slope can divide a slope into two independent grade sections (the upper section and the bottom section). The stress distributions of two sections are similar. The big platform can reduce the stress concentration at the toe of slope and change the shear stress of slopes and the distribution of shear stress, which can improve the stability of slope. According to the research on the current situation of Yan-Huang Expressway, the concept of 'wide platform with steep slope' is applied to deal with the treatment on the landslide of deeply excavated slope. For example, it can achieve good effects that setting a big platform with the width of 8–12 m at the middle of slope.

Collapsible loess cutting slope can result in slope landslide, scour and other engineering accidents frequently due to poor temporary drainage. In this revision, the requirements on temporary drainage during the process of engineering has been put forward.

7.10.5 This revision refers to the current concept of the Department of Construction and Railways on the normalized solution of collapsible loess subgrade. It summarized the practical experience of provinces (autonomous regions, municipalities) on building expressways in loess regions. According to the significance of and structural features on fabric and embankment, as well as the danger after being soaked and the difficulty of repair, the minimal treatment depth has been defined in Table 7.10.5.1, integrating the characteristics of highway engineering.

Collapsible loess is widely distributed. Therefore, the cost of treatment on it is relatively high. In this revision, common treatment on collapsible loess and the concerning application range has been put forward (see in Table 7.19). When selecting proper treatment on collapsible loess subgrade, treatment plans need to be in line with the requirements of motorways on subgrade deformation, the thickness and natures of collapsible loess, applicability of treatment, engineering conditions, material sources and etc. Meanwhile, the concerning plans need to also be made after comprehensive analysis on technical and economic conditions.

The treatments on loess pitfall and the concerning application are as follow: backfilling compaction is used to deal with overt holes; open cut on backfilling compaction to shallow hidden holes; supporting backfilling compaction to deep hidden holes; pumped sand to small and straight hidden holes; mud grouting to big and deep hidden holes. To avoid new loess pitfalls, drainage measures need to be taken on the surface water that flows into pitfalls, so as to prevent rainwater infiltration.

7.10.6 The design of high loess embankment and deep cutting and construction shall be conducted concurrently and modified in time. The problems encountered in the process of construction shall be responded to designers by constructors in time. Designers shall check and modify the plans in time and then send the revised plan to the constructors. In the process of engineering, the deformation situation and stability of substructure, road surface and slope shall be monitored. Any potential risk in engineer shall be informed in time, so as to put forward adjustment plan in time and engineering plan.

7.11 Substructure in saline soil areas

7.11.1 Salt heaving, turn swelling, melt sinking and other diseases are frequently happened on the roads in saline soil areas due to the comprehensive effects of surface water, groundwater, temperature and dynamic load. These diseases can take a toll on the construction, management and maintenance of roads.

The diseases of saline soil substructure are caused by the interaction of salt, water and temperature. Saline is the root cause of melt sinking, salt heaving and other diseases. To prevent diseases, the condition of substructure and foundation should be improved from the perspectives of salt, water and temperature and etc. The saltiness of substructure fillers need to be limited in a reasonable range. The salt prevention, membrane barrier and drainage design of substructure and foundation need to be attached great importance.

The settlement of saline soil substructure is the key point of the road reconstruction project in saline soil areas. The focal points of foundation reconstruction are: (1) the root cause of existing substructure diseases; (2) the utilization plan of existing substructure; (3) the settlement on existing diseases of salinized soil substructure. When designing, the utilization ratio of existing substructure shall be strengthened. The comparison of reconstruction plan is also conducted from the perspectives of technology and economy. In these way, a reasonable reconstruction plan of substructure can be formulated, thereby eradicating all old road diseases.

7.11.2 The methods of classifying saline soil are quite a few. According to the formation process of saline soil, it may be classified into contemporary salt accumulation saline soil, residual saline soil, alkalized saline soil. According to the degree of salinity, it may be classified into weak saline soil, medium saline soil, strong saline soil and over saline soil. According to the nature of salinity, it may be classified into chlorine saline soil, sub chlorine saline soil, sulfurous saline soil, sulphate salty soil and carbonate-saline soil.

The classification of saline soil engineering is according to the content ratio of chloride, sulfate ions, bicarbonate ions and bicarbonate ions. Such a classification method has been applied for a long period, confirming to the reality to some degree. However, today's classification system still has two salient problems: (1) the classification system of coarse saline soil and test preparation grain size still need to be improved. (2) the classification based on degree of salinity merely cannot well correspond to the actual salt expandability of subgrade. As demonstrated by research and engineering, the main factors that decides the characteristics of saline soil engineering include the granulometric composition of saline soil, saltiness characteristic, moisture content, temperature and etc. At present, the related research has still been conducted. This classification still uses previous methods of classifying saline soil.

7.11.3 The saline soil foundation needs to be evaluated based on the salt expansion and collapse behavior of soil.

According to the results of tests and researches, salt expansion can accurately reflect the damage degree of roads. When salt expansion ratio is below to 1%, road surface is flat without fissures, neither salt heaving damages (non-salt expandability); when salt expansion rate is at the range of 1%-3%, few fissures can be saw at the surface of road and weak salt heaving happened (weak salt expandability); when salt expansion rate is at the range of 3%-6%, obvious fissures can be saw at the surface of roads with salt heaving happening (medium salt expandability). Therefore, this standard revision will adapt salt expansion rate as the evaluation indicator on salt expandability.

The primary cause of salt heaving in saline soil is the sodium sulfate in soil. With the change of temperature, sodium sulfate can absorb water and crystalize with size expansion. Transforming from sodium sulfate anhydrous into $Na_2SO_4 \cdot 10H_2O$ can bring about 3.1 times of expansion on size. The result of site observation demonstrates that salt expansion will happened when the content rate of sodium sulfate in soil is over 0.5% and the temperature of soil mass is below to 5 °C; obvious salt heaving can be observed at the surface of road if the content rate of sodium sulfate in soil is over 1.2%. The higher the content rate of sodium sulfate in soil, the stronger the salt heaving capacity. The corresponding relation between the content rate of sodium sulfate and the rate of salt heaving is shown in Table 7.6. Salt expandability can be

Table 7.6 The relationship between salt expansion rate and sodium sulfate content

Salt expansion rate η(%)	$\eta < 1$	$1 < \eta \le 3$	$3 < \eta \le 6$	$\eta > 6$
Sodium sulfate content Z(%)	$Z \le 0.5$	$0.5 \le Z \le 1.5$	$1.5 \le Z \le 3.5$	$Z \ge 3.5$

Table 7.7 The classification table of collapsibility level

Collapse level	Collapse settlement ΔS (cm)	Collapse level	Collapse settlement ΔS (cm)
Not collapse	$\Delta S < 7$	II	$15 < \Delta S \le 40$
I	$7 < \Delta S \le 15$	III	$\Delta S > 40$

evaluated by testing the content rate of sodium sulfate in soil for lacking conditions that can test salt heaving rate.

The melt sinking of saline soil includes melt sinking distortion and water creep distortion. The former one refers to the subsidence caused by the damage of soil structure when hydraulic gradient is low without seepage and the salt in soil dissolved entirely and partly. Water creep distortion refers to the distortion caused by the enlargement of soil voids under the influence of seepage, which can bring away the salt and part of solid particles then bring about water creep. When evaluating the collapsibility of saline soil, the melt sinking factor δ is adapted as an indicator. According to China's *The Construction Standards in Saline Soil Areas* (SY/T 0317—1997), non-collapsible saline soil is with $\delta < 0.01$. In terms of evaluating saline soil, whether the soil belongs to collapsible soil or not should be defined at first according to the melt sinking factor δ. Subsequently, collapse degree ΔS needs to be calculated. Then the classification of collapse degree shall be finished according to the Table 7.7.

According to the features of roads, this standard revision puts forward the control standards of collapse degree for the foundation of road at all levels. The standards are shown in the Table 7.11.3.2.

7.11.4 The settlement on saline soil subgrade mainly aims at controlling the salt content of foundations, so as to lower the influence of the salt heaving and collapse of foundations on subgrades. Salinized weak foundation refers to the content of diffluent salt in muck, mucky soil and other high compressive soil and surface layer of foundation is close to or over the standard values. Such a subgrade has the characteristics of both soft soil and saline soil. Saltness has great influence on the mechanical property of soft soil. The settlement on foundation can be classified into surface settlement, shallow settlement and deep settlement. Settlement design needs to be made according to the salinization degree of surface and the thickness of soft weak soil layer. The settlement plan of foundation should not be influenced by diffluent salt and be corrosion proof.

7.11.6 When defining the height of embankment in saline soil area, the rising height of capillary water, salt heaving depth, groundwater depth and the height of surface water and other factors should be taken into consideration, so as to prevent groundwater and surface water influencing foundation workplace, as well as salt accumulation causing slat heaving of substructure and collapse. According to engineering experience and research results, Table 7.24 lists the minimal height of embankment without setting insulating layer.

7.11.7 As a substructure filler, saline soil has a close relation to the stability of substructure. Previously, substructures were frequently damaged for without strict requirement on fillers in the process of engineering. Different salt-nesses and property of fillers have great difference on the stability of substructure. In different climatic regions or under different hydrogeological conditions, the usability of saline soil as substructure fillers can also be different. Meanwhile, the filler of different levels of embankment (roadbed, upper embankment, bottom embankment) can have different effects on the stability of substructure. Roads at different levels also have difference requirement on the stability and durability of substructure.

Therefore, as a substructure filler, saline soil can be controlled separately according to the following classification: The salt properties of chloride saline soil, sub-chlorine saline soil, sulfate saline soil and sulphurous acid saline soil;road levels; layers;soil textures. When designing, the saltness of fillers need to be strictly controlled in the following conditions: (1) the filler at the range of the roadbed at high level; (2) sulfate saline soil and sulfite salty soil; (3) silty soil fillers and clay fillers above insulating layer; (4) the influence by capillary water or groundwater.

7.11.9 Setting an insulating layer at the middle of substructure can effectively prevent substructure salinization caused by the rise of capillary, as well as the substructure diseases of saline soil.

According to the water permeability of materials, insulating layers can be classified into permeable insulating layer and impermeable insulating layer. The materials of permeable insulating layer are gravels, grits and sands. The materials of impermeable insulating layer are geosynthetics (geomembrane, composite geomembrane, drainable waterproofing plate and etc.) and asphaltic sand stone. Though geosynthetics has the natures of water-proof, salt-proof and easy for engineering, it is not beneficial to the effusion of water and vapor at the middle of subgrade. Under its coating, it is easy for water and slat to accumulate, which will soften the soil layer under the coating, thereby causing new non-homogeneous deformation of substructure. To prevent the accumulation of salt and water, geosynthetics should use the geomembrane (two cloths and one coating) and compound waterproofing drainage board that meet requirements.

7.12 Substructure in perennially forzen area

7.12.1 Permafrost refers to the soil (rock) maintains frozen for two years or longer. According to ice content, it can be divided into low ice frozen soil, icy frozen soil, rich ice frozen soil, saturated ice frozen soil and ice layer containing soil. Among them, rich ice frozen soil, saturated ice frozen soil and ice layer containing soil are collectively called frozen soil with high ice content. According to the annual average ground temperatures, permafrost can be divided into low temperature frozen soil (annual average ground temperatures $\leq$ 1.5 °C) and high temperature frozen soil (annual average ground temperatures > 1.5 °C).According to ice content and annual average ground temperatures, it can be classified into stable frozen soil, basically stable frozen soil and unstable frozen soil.

4 The design of substructure in permafrost area need to be in line with frozen soil type and annual average ground temperatures, as well as under the guidance of the principal that protect frozen soil, control melting speed and allow melting.

The design principles of protecting frozen soil refers to adapt effective engineering measures to protect the existence conditions of permafrost soil, as well as maintain its original upper limit and stability. The substructure design of frozen soil sections with low temperature needs to also follow the design principles of protecting frozen soil.

The design principles of controlling melting speed refer to controlling the melting speed of permafrost soil by engineering measures, that is to say, melting within certain range is allowed. However, this range must be acceptable. In terms of bituminous pavement, within the design life of pavement, the decrease on the artificial upper limit of underlying permafrost soil can cause the thaw collapse of permafrost foundation. This collapse shall be kept within an allowable range of design. The substructure design of frozen soil sections with high temperature may adapt the principle of controlling melting speed.

The design principle of allowing melting is settling permafrost soil as common substructure and allowing melting, even damaging the existence conditions of frozen soil or accelerate its melting in advance to create conditions for construction. Substructure in low ice frozen soil and ice frozen soil section may be designed under the principle of allowing melting.

5 Rich ice frozen soil, saturated ice frozen soil and ice layer containing soil, as well as other unfavorable geological conditions frequently have serious substructure diseases resulted from sudden, severer, lasting, uneven and large thaw collapse or ice damage. Therefore, special design on these geological conditions shall be made according to the ground temperatures and types of frozen soil.

7.12.3 Substructure in frozen soil with high ice content section needs to adapt embankment plan as much as possible. When using cutting, replacement needs to be conducted on the frozen soil within the range of roadbed. The thickness of replacement needs to be decided after thermomechanical calculation according to the thermodynamic properties of materials.

The nature upper limits of permafrost areas are all below to 5 m in common circumstances. The super permafrost water of permafrost areas is relatively rich. Besides, after being excavated, permafrost soil will melt after the thermal effects of solar radiation and atmosphere, which is the main cause of secondary diseases after highways being open to traffic. Therefore, in case of solifluction, slope slide and other diseases happened, certain engineering measures on insulation and refrigeration needs to be taken to ensure the thermal stability of cut slope and avoid the damage of super permafrost water on substructure.

7.12.4 Among low temperature frozen soil, rich ice frozen soil, saturated ice frozen soil and ice layer containing soil are all called as low temperature frozen soil with high ice content.

The substructure design of low temperature frozen soil with high ice content section needs to ensure that the upper limit of permafrost soil will not drop. The minimum height of fill for embankment needs to meet the requirement of prevent melting, frost heaving and frost boil, as well as avoid foundation too high and heat-absorbing surface of side slope enlarged, which can cause the obvious drop of the upper limit of frozen soil under side slope on sunny side, and uneven distortion. The minimum height of fill of foundation and the critical altitude of substructure design shall be calculated in line with the related stipulation in *Technical Details on Permafrost Soil* (JTG/T D31-04—2012). When the calculated minimal height is lower than 1.5 m, the minimal height of embankment is not considered as lower than 1.5 m in most cases.

7.12.5 Among high temperature frozen soil, rich ice frozen soil, saturated ice frozen soil and ice layer containing soil are all called as high temperature frozen soil with high ice content.

In high temperature frozen soil with high ice content section, the critical altitude is no longer a control indictor of substructure design. However, maintain certain height of substructure is premised by various protecting engineering measure, as well as improving the anti-disaster capacity of substructure and protecting frozen soil environment.

The substructure of high temperature frozen soil with high ice content section, based on certain design altitude of substructure, needs to use substructure plans with special structures, including XPS board insulation substructure, ballast (block-stone) embankment, thermal pipe foundation, vent pipe foundation and thermal pipe- XPS board insulation substructure and etc. As such, the stability of substructure can be ensured. Substructure with special structures needs to be designed and calculated according to the concerning stipulation in *Technical Details for the Design and Construction of the Highways in Permafrost Areas* (JTG/T D31-04—2012).

7.12.7 Designing the substructures in permafrost areas needs to take the seepage in foundation into consideration. Due to poor drainage, the water accumulation at the toe of slope will cause the drop of the upper limit of permafrost soil, thereby causing the collapse of embankment. Therefore, measures need to be taken to drain out surface water, so as to avoid water accumulation around the toe of embankment, as well as seepage in foundation.

The cross dimensions of drain need to have enough discharge capacity to prevent silting and make it easier to clean up. Generally, bottom width should not be lower than 0.6 m, and side slope cannot be steeper than 1:1. If necessary, turf, clay, or dry-laid rubble (setting waterproof geotextile) can be selected to conduct reinforcement.

In the sections with thick underground ice and permafrost swamp, water retaining dike is favorable than drainage ditch, as it will not damage ground vegetation, which can help to protect permafrost soil. In the sections with drop height and without the conditions of digging deep drainage ditch, both water retaining dike and drainage ditch can be taken into consideration. However, the distance between the side of drainage ditch and the toe of water retaining dike is not smaller than 1 m usually, so as to ensure the completeness and stability of drainage system.

Table 7.8 Desert highway primary division

Code	The name of Desert area	Dryness	The ratio of hot and cold (January and July average temperature)	The name of the desert
I	Semi humid and frigid desert area	1–1.2	> 10	Hulun Buir sandy land, Nenjiang sandy land
II	Semi humid and cold desert area	1.2–2	0.7–0.8	Horqin sandy land, Hunshandake Sandy Land
III	Semi arid warm desert area	1.5–2	0.4–0.50	Maowusu Desert, The eastern part of Kubuqi Desert
IV	Arid warm desert area	8–16	0.4–0.50	In the western part of Kubuqi desert, Ulan Buh Desert, Tengger Desert, Badan Jilin desert
V	Extremely arid, cold desert region	16–32	0.8–1.0	Qaidam, Gonghe desert and Northern Tibet scattered desert
VI	Arid, warm and cold desert area	4–8	0.70–0.8	Gurbantünggüt Desert
VII	Extremely arid, hot desert area	>32	<0.40	The Taklimakan Desert, desert Kum DAG

7.13 Substructure in wind drift sand area

7.13.1 Wind drift sand area is a general term of desert and sandy land. According to the degree of aridity and the ratio of heat to cold, China's desert is divided into the following seven primary areas. See Table 7.8.

The main characteristics of sandy areas are: dry climate, small rainfall, large temperature difference, severe cold and hot changes, large wind and sandy soil; more soluble salt in soil; sparse and low vegetation.

The main diseases of highway substructure in sand blown area are sand burial and wind erosion. Sand burial mainly has two kinds: One kind is when the sand wind flow through the substructure, the wind speed weakens causes the sand accumulation and burying the substructure. Another is the movement of the dunes, which leads to the burial of the substructure. Wind erosion is the sand or soil particles which directly eroded the substructure slope with sand drift, it results in the decrease of width and height of substructure, and the hollow and collapse of slope. The extent of wind erosion is related to wind force, wind direction, cross section form of substructure, composition of filling and protection measures.

In order to prevent substructure sand damage, according to the characteristics of aeolian geomorphology, aeolian sand movement characteristics, combined with the terrain, wind force, wind direction, route and angle conditions, it should reasonably determine the position and cross section of substructure, and make engineering and plant protection of substructure.

The Ditch or culvert in arid and extremely arid desert areas are prone to be buried by wind and sand due to the lack of precipitation. Thus, it generally doesn't considerate to set substructure ditch, culvert or other drainage facilities. But in the semi humid and semi-arid sandy area, it should set the ditches and other necessary drainage facilities, according to the rainfall condition, cross sections are usually in the form of wide shallow streamline to make sand flow smoothly through the side ditch drainage facilities and reduce the accumulated sand in drainage facilities.

7.13.2 The wind speed on the embankment increases with the increase of embankment height, the wind speed on the embankment increases with the increase of embankment height, and the wind speed of the higher slope is larger, and it is not easy to produce sand burial hazard. But when the embankment is too high, the wind erosion on the slope will increase. Therefore, the embankment should have a reasonable height. At medium height dunes and sand ridges, the embankment should be 0.5–1 m higher than the average height of the dunes; when the route is parallel to the wind direction, the wind speed at the top of the substructure is larger than that on the sand on both sides of the embankment, and the sand will not accumulate in general; no fill, dig or near zero fill and dig substructure, regardless of the relationship between the route and the wind direction, are easy to accumulate sand. In tall composite sand ridges or composite sand dune chain section, height of substructure filled should be little more than or close to the equilibrium strategy of excavation.

To facilitate the smooth flow of sand drift through the substructure and reduce the harm of sediment accumulation, the embankment cross section usually adopts streamlined and slow slope. For the benefit of sediment transport, the standard needs to be adopted in the form of separated and slow slope substructure, without raised curb stone or barrier. The different heights of the substructure side slope rate are given in Table 7.27. When the route and the main direction angle is small, steep slope ratio is considered. When it takes protective measures on the road side, the slopes may also be moderately steep. Aeolian sand is widely distributed in desert area. Its grain size is mainly 0.25–0.075 mm, the content of silt and clay is little, the plasticity index is close to zero, and the density is 2.65–2.7 g/cm^3. Aeolian sand is a good road building material, but it is difficult to construct. In the design, it should make full use of it as substructure material. In order to improve the overall commander bearing capacity and convenient construction, generally, paving geotextile is used for sand fixing, and vibration dry compaction technology is adopted.

7.13.3 The road cutting are easier to accumulate sand than embankment. When road cutting is short and shallow, sand will easily be brought to the outside under the action of ditch along wind. When road cutting is too long, it is easy to form their weak wind area and lead to sand accumulate. When the angle between route and wind direction is orthogonal, cutting easily have sand in different degrees, deep cutting result to more serious sand accumulation. Therefore, it is necessary to avoid using a longer cut as far as possible during the design.

In general, when the cutting slope is steeper than 1:4, accumulated sand condition in road cutting will be more serious; when the slope ratio is lower than 1:4, slower slope will make wind flow through the cutting more smoothly and accumulating sand in cutting will be not easy. Table 7.28 lists different slope cutting depth ratio, according to the route and the main wind angle by slope ratio, when the angle is small, the slope rate is steep.

7.13.4 For the substructure in the wind and sand region, it is necessary to protect the road and shoulder, slope face and the range of 20–30 m from the top of the cutting ramp and the sediment platform, protection design should follow the local conditions, the principle of local materials, selection of protection materials reasonably and economically. The requirements for various types of protection are shown in Table 7.9. Polymer geogrid and geocell material has the anti-aging trial instead of firewood material for protection of substructure sandy area with good results.

7.13.6 The roadway sand control project is divided into fixed sand, blocking sand and transportation sand, and combining engineering and plant measures. In order to make various kinds of preventive measures of economic efficiency, the formation of comprehensive protection system, it is necessary to carry out the general layout and engineering design, the main contents include: the overall layout and design, cross section of substructure design, substructure protection and roadside sand control engineering design.

Semi humid and semi-arid sandy areas (No. I, II, III) belong to grassland or Desert sandy land. The water vapor condition is better, rainfall is 200–400 mm, which is beneficial to plant growth. Sand control project should be based on plant sand control and restoration of local ecology.

Arid desert region (No. IV, VI) belongs to the arid desert, The water vapor condition is normal, rainfall is 100–250 mm, plant growth is difficult. Generally, use the combination and fixation method of engineering and plant, gradually transition, and finally achieve plant control. For the place with the ground water where has an Irrigation condition, it must first adopt the plant to control the sand, build the sand control belt.

In the extremely arid desert area (V, VII area), the sand drift is seriously damaged. The water vapor condition is very poor, the rainfall is less than 100 mm, the sand source is rich, the sandstorm flow is strong, and the dune movement is fast. In addition to the basic body protection, it also need to establish the perfect system of prevention and protection zone in the roadside, including flat belt, sand belt, vegetation protection belt etc., only by using the integrated protection system of blocking, fixing, transportation and engineering measures, the sand damage can be effectively suppressed; for those sections which are not seriously damaged by wind sand flow, only partial sporadic sand damage should be controlled, and sediment transport measures should be the main method.

7.13.7 1 The action of sand blocking measures is to intercept sand drift and limit the accretion movement. Generally, barrier barriers can be divided into five types: wall, dike, grid, belt and windbreak. They are suitable for sandy areas with extremely rich sand resources. They need to be located over 80 m outside the windward side of the substructure, and are generally located at the top of the dunes. The higher the sand barrier is, the larger the spacing is, and the better the sand blocking effect is when it is orthogonal to the dominant wind direction. Vertical sand barrier is an effective

Table 7.9 Types of substructure protection

Protection types		Material	Protective thickness (cm)		Scope of application
Classification	Form		Shoulder	Side slope	
Firewood protection	Layering	Wheat straw, reed, artemisia etc.	5–10	5–10	Classes-2 and below 2 highways
	Paving plant beam or flower block	Wigs, reeds, achnatherum etc.	5–10	5–10	
	Paving or stacking sward	Sward	Tiled 10–15	Tiled 10–15	
Soil protection	Spreading Paving clay	The clay which plasticity index is greater than 10	10–15	5–10	
	Paving salt cover	Rock salt cover	10–15	5–10	
Gravel, pebbles, rubble protection	Paving gravel, scree	Pebble stone	10–15	5–10	Motorway and Class-1 highway
	Cut stone lattice Gravel, pebbles, rubble	Gravel, pebbles, rubble	10–15	5–7	
Cement concrete protection	Prefabricated chunk	Precast slab, block, Hollow lattice net	5–10	10–30	
Asphalt protection	Paving asphalt sands	Asphaltic sands	5	5	
	Spraying asphalt emulsifier	Low grade asphalt emulsifier	2	2	
Geosynthetics protection	Paving geogrid	Polymerizate geotechnical materials	–		
	Using geocell				
Plant protection	Cultivate	Sward	–		

sand fixing method, which has better sand blocking function. When the conditions are suitable, plants and shrubs can be planted first to form plant sand barriers.

2 The role of sand fixing measures is to fix the local sand, improve the surface roughness and weaken the wind speed. Engineering sand fixation is the formation of lattice or banded sand barriers and sand fixation using various materials, gravel, clay tile or using other materials such as sand or sand material by chemical synthesis. Vegetation is the basic measure to control the harm of the sand. It not only can reduce wind speed, weaken and restrain sand flow activities, but also can consolidation of sand, which is conducive to the accumulation of organic matter. It also changes sand

properties so that the sand flow tends to be fixed. This method is preferred when qualified.

When the width of sand fixing belt is too large, the cost will be increased. While if it is too narrow, it will reduce the effect of sand control and result in sand ridges and dunes, which will form sand buried. As for the width of sand fixing belt, it is stipulated in the regulation that the main wind direction should be 60–200 m on the windward side. The leeward side of the one-way wind is not definitely need to be provided, and the width of the leeward side of the reverse wind shall not be less than 50 m.

The design should be combined with the local wind direction, wind sand damage degree and sand prevention experience to set up.

3 The role of sediment transport measures is to increasing the wind force or change the surface properties, so that the quicksand can pass through the substructure without accumulation. The method mainly includes streamlined, slow slope, wind embankment, shallow trough, wind board, etc.

Flat belt is an important part of roadside Sand Preventing system. Especially in the flow and the dominant wind angle of 45°–90° of the flow and semi-fixed dune area, on both sides of the substructure 20–30 m range of flat belt, is conducive to smooth flow through the substructure to reduce or avoid the sand damage.

4 The forest belt should be combined with the grass, and the belt will be arranged on both sides of the substructure. The formation of the system of comprehensive protective zone, needs certain conditions, has difficult management, and need a long time. It should be combined with local sand-control programme and rely on local people enthusiasm and agriculture and forestry department work closely.

5 In order to protect the natural growth of the sand belt, it is necessary to set up the vegetation protection belt outside the sand belt. Typical dune area, the width of the vegetation protection belt usually adopted is as follows: The flow dune location, the windward side 400–600 m, the leeward side of 200–300 m; semi-fixed dune location, the windward side 300–400 m, the leeward side 100–200 m; in fixed dune area, the vegetation protection belt is set up directly on both sides of the substructure. The wind side is 300–500 m and the leeward side is 100-200 m.

7.14 Substructure in snow disaster area

7.14.1, 7.14.2 Highway snow disaster is mainly divided into windblown snow and avalanches. Windblown snow refers to, when the wind reaches a certain intensity (4–5 m/s) during snowing or after snow, snow particles move with the wind and eventually forms a snow flow. The main hazards are snow blocking vehicles and snow blocking the line of sight. Avalanche refers to, under the impact of gravity, the hillside snow collapse, bury highway, block the passage.

For the hazard area of wind blowing snow, it is important to investigate the topography, ground features and vegetation in the snow-collecting process, to determine the amount of snow-drift, the winter wind and wind direction and its frequency and duration, snowfall, snow depth, winter temperature, freezing and thawing time, type of wind-blown snow and its harm degree, etc. In avalanche area, it is important to investigate the range of avalanche distribution, the location of crack point and the frequency of damage.

The substructure design in the snow-disaster area should follow the principle of 'Prevention is more important than cure'. On the basis of investigating and collecting the natural geography, climate, disaster, snow, dominant wind direction and ecological condition of the corridor belt, it needs to be reasonable to choose the route plan through the snow harm area, to avoid serious snow damage. It also needs to try to set layout line on open ground, platform, ridge, long hillock and other terrain. And it prefer to choose sunny slopes, Windward Hillside and lay the substructure in these areas with the least amount of snow and lighter snow damage.

7.14.4 The formation and harmfulness of wind-blown snow are closely related to terrain features and substructure cross-section. Flat open terrain facilitates smooth passage of airflow; when the direction of the route and the wind direction of the snow flow near the parallel, the road will form a smooth flow field, can reduce the snow on the road; low-fill cutting easily produces snow; the embankment with the appropriate height can ensure the smooth passage of snow flow; the streamlined or gentle side slope cross-section form is most advantageous to prevent accumulation of snow. There are obstacles in the 20 m range on both sides of the substructure, which easily cause snow on the road. Therefore, according to wind and snow conditions and terrain conditions, the cross section of substructure should be chosen reasonably, during the design.

7.14.5 The reasonable determination of embankment height and side slope rate is an important measure to prevent and mitigate substructure disease in windblown snow area, during the substructure design.

The embankment in the windblown snow area needs to be higher than the local maximum snow depth. Its minimum height may be determined by the nearby terrain, wind, wind direction and snowfall. In this revised code, according to the investigation of substructure disease in the windblown snow area, it is proposed that the minimum height of embankment is 'the maximum thickness of the local snowfall in 50 years, plus safety height'.

Slow down the side slope rate is an effective method to prevent the harm of snow drift. Grassland, agricultural and pastoral embankment slope rate using 1:3, wasteland Gobi embankment slope rate using 1:4, can ensure that the road does not sediment windblown snow, but also is beneficial to vegetation growth.

7.14.6 In mountainous excavation section, the mutual interference between main wind direction and local vortex cyclone is easy to produce snow in the weak wind area. Thus, it needs to avoid the deep cutting in design. The cutting usually adopts a gentle slope open type cross section which facilitate the snow circulation (the top of the side on the windward side is higher than the top on the leeward side) and set up a snow platform to make the road surface no snow or less snow.

When the route direction is parallel to the dominant wind direction or the sharp angle intersection, If the vertical curve slope changing points in inside of cutting, it easily forms the weak wind area in the vertical curve Leeward section and causes the road surface snow. Hillside cutting, especially when the angle between its route and the main wind direction is greater than thirty degrees, is easy to produce the windblown snow accumulation. The amount of snow in the cutting and the snow area are proportional to the snow area of the hillside on the windward side of the cutting. when the snow disaster is lighter, the safe height take the lower limit value, otherwise take the upper limit value.

7.14.7 The half fill and half excavating substructure the alignment of which is an acute angle with the wind direction is in the wind flow of snow, but the excavation side of the mountain will cause the snow flow around, the formation of snow in the excavation side, blocking traffic. Usually the measure of widening the excavation width, and width should also be increased at least to 2 m.

For the half fill and half excavating substructure the alignment of which is 45° to 90° to the wind direction, if the filling side of it is suffered wind erosion, it will lead to the wind and snow flow maneuver on the side of the excavation and accumulative snow on the road surface; if the excavating side of it is suffered wind erosion, it will produce leeward piles of snow on the excavated side slopes. Therefore, the substructure usually uses the gentle slope rate of open cross section, in order to ensure that the wind and snow flow smoothly through and the road no snow.

7.14.8 Highway wind-blown snow control project may be divided into engineering management and plant management, engineering management mainly using the stability, resistance, guidance three kinds of treatment measures.

1 The protective effect of Snow protection plantation belt is closely related to the belt structure of the forest (cross-section form, height, and ventilation degree), and the snow protection plantation needs to be planted closely, so that the wind and snow flow is not easy to pass. And it should be advisable to use arbor and shrub mixed forest.

2 Snow Prevention fence is widely used at home and abroad, the main factors that affect its protective effect are the ventilation degree, plank structure, height and terrain of the snow Prevention fence. The snow prevention fence should be set on the windward side of the flat terrain road section. The calculation of barrier voids of snow prevention fence can be referred to *Handbook of Highway Substructure Design.*

3 Closed-type and ventilated-type downward wind baffler are effective measures to prevent and control the storm. In comparison, the latter has the advantage of reducing the height of the snow in the front and less surface material of wind baffler.

The upwind half cutting which has a large angle between route and dominant wind direction, the embankment with angle 30 degrees above, the leeward half cutting with angle 40 degrees below commonly use lateral wind baffler (Fig. 7.2). The upwind half cutting with lager angle, when the leeward slope is steep or the snow storage field is narrow and it is difficult to set other protective measures, is usually set the open type lateral wind baffler (Fig. 7.3). In the turn around the flow and other sections, it can also use the treatment methods which combine the downward wind baffler and lateral wind baffler combination.

7.14.10 Highway avalanche prevention and control projects include horizontal steps, snow fences, snow fences, snow forest, mound and wedge, guide snow dike, snow wall, snow-proof corridor, etc. During the design, it need to be suitable for local conditions and make reasonable selection, according to the terrain conditions

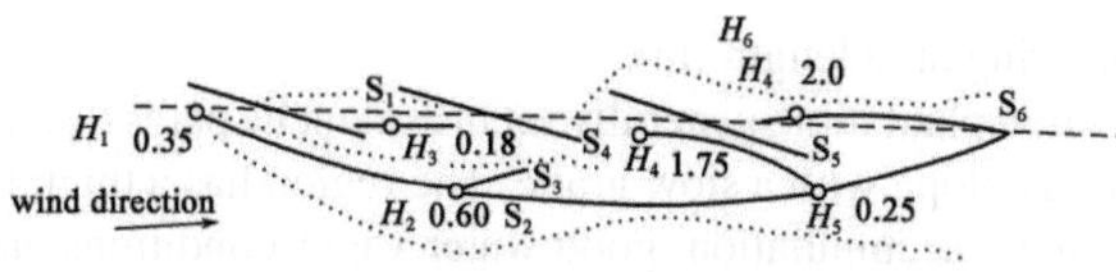

Fig. 7.2 Enclosed side draft wind snow shape S_1......S_6-Snow ridge; H_1......H_6-deep snow(m); o-Snow depth measuring point position;......-Snow contour

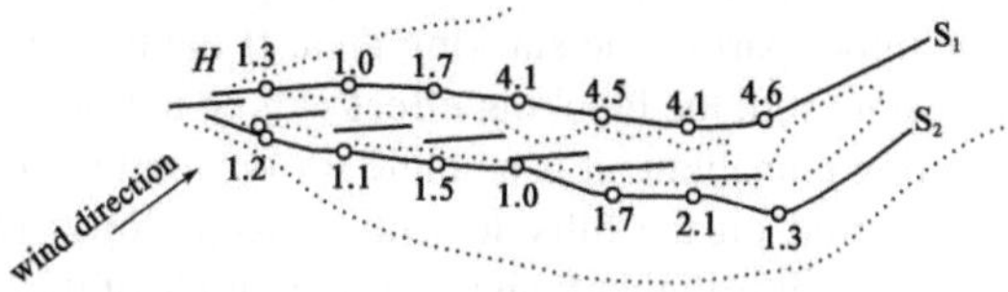

Fig. 7.3 Open side draft wind snow shape S_1......S_6- Snow ridge; H_1......H_6-deep snow(m); o-Snow depth measuring point position;......-Snow contour

and scale of avalanche, combined with the applicable conditions of various types control engineering.

1 The horizontal steps are mainly used to control small avalanches, which are usually arranged along the contour lines of slopes along the avalanche path, so as to change the micro topography of the slopes and stabilize the accumulated snow on the slopes.

2 The stable snow fence is used to prevent avalanche on slopes with steep slope, thin soil layer, poor permeability and unfavorable excavation of horizontal steps. It usually is set along the hillside contour and Avalanche path. Where conditions permit, the snow fence is used alternately with the horizontal steps. The snow edge mentioned in the provisions refers to a thick hanging snow body which is formed by the wind-blown snow in inside watershed. When its weight increases to a certain extent, it will break down and form an avalanche. The snow breaking point described in the clause refers to the highest position of the snow-collecting area of the windblown snow.

3 The block snow fence generally uses the wood or the metal and other material, usually is set in the avalanche motion path area of slow grade slop. The fence needs to have higher strength, need to be buried into a certain depth and have certain stability and resist the impact of avalanche.

The maximum front attitude of the snow slid may be calculated as follow:

$$h = \frac{FH}{bL}$$

where:

h—the maximum front attitude of the snow slid (m);

F—snow gathering area (m^2);

H—maximum snow depth in snow gathering area (m);

b—channel width (m);

L—snow gathering area length (m).

4 Snow protection plantation, generally set in avalanche paths occurrence or movement area of cloudy slope with a slow grade. The region has a thick layer of humus, large avalanche snow accumulation, good water vapor conditions and is conducive to plant growth. It can be set along the contour line to block the slide of snow on the hillside. It should be advisable to set up multiple rows.

5 The mound and wedge are suitable for preventing and controlling groove avalanches with relatively slow slope and large scale, generally set in the Avalanche motion area and the upper part of the stacking area. If the height is lower than the front attitude of the snow slid, the blocking effect will be not obvious.

6 The snow directing embankment is suitable for preventing and controlling the groove avalanche occurred on the hillside with a steep slope. The height of the snow directing embankment is higher than the front attitude of the snow slid and the collapse snow body on slope surface can be exported from the substructure.

7 The function of the snow retaining wall and the snow prevention fence is basically the same. Snow-retaining walls are generally used in the avalanche area, which is rich in earth and stone materials and set in the motion or stacking area of an avalanche path. It can be divided into dry-masonry (slurry) stone or reinforced concrete. The wall needs to have high strength and stability to resist the impact of avalanche.

8 Anti-snow corridor, tunnels and other shelters are the most effective means to prevent avalanches. Because of its large engineering and high cost, it is only used when avalanches on high grade highway are frequent and other measures are difficult to adopt.

7.15 Substructure in ice drift section

7.15.1 The ice drift is divided into hillside ice drift and valley ice drift. Hillside ice drift is formatted by exposed freezing ground water from the hill slope or the slope excavation. The valley ice drift is formed by freezing water along the gully and ice melt. It is important to investigate the amount of ice flow, the flow range, the type of ice drift and the local control experience of various water sources during the cold season.

Water is the ice source of ice drift. The key points to prevent and control of the substructure in the ice drift section are harnessing water and preventing ice, so as to make the water flow smoothly pass through the substructure or change the direction of water flow, to maintain the substructure is not affected, in order to cure or reduce the damage of ice drift to highway. The comprehensive measures to prevent and control ice drift include: avoiding, adopting embankment, improving the clearance or deepening of the trench, adopting the measures of excluding groundwater or preventing ice. The design needs to be proved by techno-economic, and the measures of treating the ice drift need to be rationally determined. Due to the fact that most of the ice drift ice is deposited in winter and the water is destroyed in the spring, it need be maintained in time. And it is necessary to consider the maintenance and management of engineering during the design.

Ground water ice drift in micro-hilly area will form a wide and long ice flow. The use of heat preservation ditch often does not work. Usually, the water interception ditch is used to truncate the ground water, and the water flow is introduced into nearby bridges or low-lying areas. It usually uses a set of water open to cut off the dive, the water flow into the vicinity of the bridge or the lower place. The ice-retaining dam or ice storage tank also can be arranged under the ditch. If the bridge is introduced, it is necessary to consider whether its aperture and clearance meet the needs of spring-thaw season drainage and silt-discharging ice.

The different treatment methods are different depending on the soil quality in the hilly area. If the ground water seeps from the poor quality slope plot soil and the soil is saturated, after engineering disturbance, the ice drift usually is not the main disease, but slope slide and landslide are the main disease, Thus, it is necessary to take out variety of water from ground surface and soil, increase the stability of the slope.

For the ice drift produced by seepage water in the rock fissures, If the seepage quantity is small, it can enlarge and deepen side. If the seepage quantity is big, it can be deal with the heat preservation retaining wall and other measure.

For the ice drift formed by stream (ditch) water, when the channel is shallow, the substructure is lower, the bridge diameter is small, the culvert bottom slope rate is slow and the ditch mouth has alluvial heap, the ice drift will flow on the road. The saliva-flow ice needs to be governed by the method of blocking (blocking dam), guiding (guiding dam), and storage (ice storage tank). Firstly, it need to check whether the span and clearance of the downstream bridge and culvert meet the height requirement of the formation of the silt, and then build the ice guide dam from a certain position of the groove.

7.15.2 The prevention and treatment of saliva-flow ice needs to be carried out from the substructure, by improving the height of substructure, combining with the terrain condition and the bridge and nearby structure, to cure or reduce the harm of the ice drift to the highway.

To improve substructure height and choose gravel soil with good water stability as substructure filler, not only can avoid the ice drift on the road, but also to prevent the substructure softening and boiling after the melting of the ice. These are simple, economical and effective measures to prevent ice drift to harm the substructure. It is suitable for the prevention and cure of the ice drift with not too much gather amount. The height of the substructure need be 0.5 m higher than the top height of the ice drift.

It needs to make more fill and less excavation as far as possible, to avoid the construction cut off underground aquifer and form the ice drift at the excavated side slope.

7.15.3 In order to ensure the smooth flow of water and to control the river valley ice drift, the bridge design needs to be considered to deepen and clean up the river and other ancillary measures. The bridge abutment body usually chooses the cast-in-place concrete structure to prevent the frost heave damage caused by the ice drift water immersing into the human structure.

7.15.4 The retaining Ice dike (wall) is built on the base of the substructure, the lower side of the groundwater outcrop of the hillside or the upstream of the gully, bridge and culvert, in order to prevent the ice drift and reduce its spreading range. The retaining ice dike is suitable for the hillside ice drift of small amount water seepage, and it is commonly used in conjunction with the gather ice pit. The dike is usually located on the outside of the side ditch, keep the side continuous unobstructed, which is advantageous to elimination of the snow melted water thorough the culvert.

7.15.6 The outlet of ditch and blind ditch needs to be located on the steep sloping field as far as possible, and need be higher than the ground not less than 0.5 m. In the serious area of the ice drift, the water outlet of the underground drainage facility needs to take the heat preservation measure or excavate trenches with longitudinal slope rate greater than 10% to prevent the water from freezing.

7.16 Substructure in mined-out area

7.16.1 The mind-out area is a concealed world of underlying highway, the treatment of which directly influence the stability and operating safety of the substructure. Preliminary reconnaissance of highway is important then. It is also necessary to combining the problems occurred in the treatment process and make supplemental geological prospecting and adjust engineering proposal in time. Currently, the explorations of the mined-out area are mainly combination of geophysical prospecting. Drilling and ground settlement observation are the supplement. The accuracy of exploration and the reliability of the basic data for design is enhanced.

7.16.2 The methods ogre estimation of stability include: Mining condition discriminance, surface movement and deformation prediction method, ground deformation observation method, limit equilibrium method and numerical analysis method. The applicable conditions of various methods are found in Table 7.10. Estimation of stability of the mind-out area need to observe the principles of combining qualitative evaluation with quantitative calculation. Combining the methods to predict the residual deformation of the surface, evaluate the stability of mined out area and avoid the one-sidedness of a single evaluation method.

7.16.3 Table 7.30 gives the allowable value of surface deformation of mined out area in highway. It refers to relevant industry standards and engineering experience in construction, road and coal industry both at home and abroad. According to this, it determines the allowable deformation value of different grade roads to the mined out area.

When it comes to the problems of the determination of allowable value of surface inclination, surface horizontal movement and vertical curvature of highway engineering, the *Specification for Geotechnical Investigation* (GB 50,021—2001) gives the standard for building suitability of mined out areas: the horizontal scroll of the surface $\varepsilon \leq 6.0$ mm/m, surface tilt $i \leq 10.0$ mm/m, surface vertical curvature $K \leq 0.6$ mm/m^2. *Rules for coal pillar retaining and coal mining in buildings, water bodies, railways and main roadways*(State Bureau of Coal Industry,2000) specifies the allowable value of ground deformation for buildings with minor and minimal damage: he horizontal scroll of the surface $\varepsilon \leq 2.0$ mm/m, surface tilt $i \leq 3.0$ mm/m, surface vertical curvature $K \leq 0.2$ mm/m^2. Referring to the specifications and procedures above, this paper takes the minimum of them to determine the allowable stain value of ground surface shown in Table 7.30.

Table 7.10 Evaluation methods and applicable conditions of mined-out area stability

Stability evaluation method	Application condition
Mining condition distinguishing method	It may be used for the post pillar type, the irregular room and pillar type and other mined out areas which are difficult to estimate the ground subsidence and deformation
Surface movement and deformation prediction method	It may be used for mined out areas of long wall type, strip type with normal design or room and pillar type
Ground deformation monitoring method	It may be used for strip mining by strip caving method or mining with special design or room and pillar mined-out area which has regular Surface subsidence
Limit equilibrium method	It may be used for the post pillar mined-out area with narrow mining area and no moving basins on the surface
Numerical Simulation Analysis	It may be used for mined areas with deeper mining depth, larger mining range and complicated geological conditions

In China, the built and under construction highway engineering in Shanxi, Shaanxi, Henan, Hebei, Xinjiang, Inner Mongolia, Jiangsu, Hunan and other provinces all adopt this standard. After many years of tests, there are no disease or safety accidents caused by the settlement and deformation of the substructure because of the improper selection of the surface stability control standards in the mined out area highway. So this standard is appropriate.

7.16.4 The setting and the determination of coal pillar for highway protection refers to *Rules for coal pillar retaining and coal mining of buildings, water bodies, railways and main shafts* (State Bureau of Coal Industry, 2000).

When protection coal pillar is retained, enclosure is also provided outside the boundary of the protection area because of the complex overburden structure of the mined-out area and the difficulty of accurately delineate the moving angle of subsidence. The protection coal pillar is a safety measure to guarantee that the object is outside the moving boundary. The width of the retaining belt relates to the safety level of the road and the amount of coal on the road. The width of the coal pillar enclosure in this revision refers to *Rules for coal pillar retaining and coal mining of buildings, water bodies, railways and main shafts* (State Bureau of Coal Industry, 2000) and *Regulations of the State Council on highway safety protection* (2011).

7.16.5 The length and width of mined out area is determined according to relaxing standards in *Rules for coal pillar retaining and coal mining of buildings, water bodies, railways and main shafts* (State Bureau of Coal Industry, 2000).

7.16.6 The main treatment methods of highway mined-out area are excavation backfill method, grouting method, dry masonry method, mortar method, spanning method and comprehensive treatment method. The selection of treatment methods is directly related to the cost, time limit and safety of highway engineering, and it is the key to the success of the treatment. When choosing the treatment method of the mined-out area, comprehensive consideration including geological conditions and construction conditions of the site, as well as factors such as the type and structure of the proposed highway needs to be given to the features of mined out area. After the full demonstration of technical economy, the best governance plan is chosen.

2 For the abandoned but not completely collapse roadway, using non-slip casting like filling method, rubble, block masonry or grouting to support overlying strata. This method is suitable for the mined out area which is not completely collapse, with large space, small buried depth, good ventilation conditions, and with manual operation conditions.

3 Dynamic compaction mainly eliminates the settlement of mind out area or eliminates holes of roads to be built. It uses gravelly soil, detritus or granules to fill, substitute and reinforce the mined out area in order to improve the bearing capacity of foundation and reduce the uneven settlement of foundation. For the areas with thick roof soil, wide distributed low cave, we adopt dynamic consolidation (or dynamic consolidation replacement).

5 The mined out area with smaller range and difficult disposal is mainly in the following situations;

(1) The small minded out place with complex pit; whose depth is greater than or equal to 30 m;
(2) Grouting treatment of multilayer and water filled mined-out out place which is difficult to achieve the desired effect;
(3) Age-old mined out areas with difficult treatment conditions;
(4) The mined out area which the substructure can not be adopted because of the limitation of terrain and ground conditions;
(5) Tomb cave, large cellar, large cave, underground works and other non mineral mined out areas;
(6) Special mined out areas (such as underwater mined out areas).

6 For the mined out area with large coal seam mining and less than 250 m mining depth, because of the large buried depth, the serious collapse after the mining of the seam, high filling degree and relative small bottom space, it is hard to use excavation backfill, dynamic compaction, dry masonry support, slurry support and other treatment methods. To ensure the stability of the mined-out out areas and meet the requirements of surface deformation in highway mined-out area, usually we adapt the method of full filling grouting.

For the areas with over 250 m depth, the treatment plan is determined according to the combination of its mining characteristics, hydrogeology, engineering geological conditions, the degree of damage to highway engineering and other factors.

7.17 Substructure in coastal area

7.17.1 The substructure in coastal area is similar to the submerged substructure, such as riverfront and reservoir substructure, but there is something unique about it: The sea water is affected by hydrological, meteorological factors such as tides, waves, currents, typhoons and tsunamis; the basement terrain is generally inclined or flat, and there is usually a silt of varying thickness, and different types of coastal substructure have different characteristics because of their geographical environment.

(1) Besides the same driving load as the ordinary substructure, the substructure is subjected to other external force such as ocean currents and wave forces. They not only have high strength, but also have dynamic properties. In addition, in some cold regions, the substructure also is influenced by the ice. These conditions need to be fully considered in the design.
(2) Because it is often constructed on beaches or water and easily affected by wind, waves, currents and other hydrological, meteorological factors, the design needs to be closely combined with the construction conditions and geographical environment and select reasonable substructure cross section and protection form.
(3) In addition to the long-term erosion by seawater and the biological, the substructure is also affected by wet-dry change and freeze–thaw condition caused by water level fluctuation. Thus, it is required that the substructure structural materials have strong durability, water stability and corrosion resistance.
(4) Because of the above reasons, the cost of the substructure in coastal area is much higher than that of the general substructure, and the design needs to be carefully compared with other schemes such as avoiding, bridge crossing, etc.

7.17.2 The criteria for designing waves include the design of a wave recurrence period and the cumulative frequency of wave columns designed for waves. The recurrence period of design wave refers to the average year number of a particular wave cumulative frequency, which represents the long-term statistical distribution of wave elements. The standard of design wave recurrence period mainly reflects the service life and importance of buildings.

The cumulative frequency of a wave column is the percentage of the wave number that a wave element (e.g. wave) is not less than a certain number of waves.

The design wave high cumulative frequency of all kinds of buildings is based on *Code for design of Railway Special Substructure* (TB 10,035—2006) and *Manual on Highway Substructure Design* (second Edition, China Communications Press).

7.17.3 The cross section structure of substructure in coastal area is divided into sloping type and straight wall type. The sloping type cross section and the general substructure cross section form are consistent, easy to connect, and the construction is convenient, the overall stability is high, Is the common cross section form of highway substructure at present. Straight wall Embankment generally uses block stone or concrete masonry, its base often uses the riprap dark base bed, or sets the riprap on the base lateral to prevent flushes. The straight wall cross section is usually only used when there is a lack of material or other requirements for use.

7.17.4 Because the wave action of the coastal substructure slope is much larger than the substructure along the river, the slope rate of the offshore side slopes is usually slower than the inland sea side. The stability of riprap is lower than artificial armor, so the slope needs to be fattened correspondingly.

7.17.5 The upper bound of the protection range of the slope surface of the coastal substructure depends on the increase of water in front of the embankment under the action of wind and wave and the climb of the wave on the embankment slope, its lower bound is determined by the ability of the soil to resist the impact of fluctuating bottom flow lin.

In order to weaken the effect of wave on embankment, it usually takes measures to promote the elimination of waves in front of the embankment, and reduce the quantity of protection works and improve the stability of substructure slopes. The commonly used wave-eliminating measures include plant wave-eliminating, shun-dam and submerged dike to promote siltation and wave-dissipating, breakwater wave-dissipating, pressure and hydraulic breakwater waves, etc.

2 Slope protection cushion layer is the foundation of slope protection layer, the strength is not enough will lead to instability of the surface layer, so the weight and thickness of the cushion are stipulated.

3 In order to ensure the stability of slope protection foundation, it is usually necessary to set up riprap prism on the sea side. In order to facilitate the construction, the top surface of the riprap body should be high out of the construction water level. The top width and thickness of the prism are based on the relevant provisions of the *Design and Construction Regulations of the Breakwater* (JTS 154-1—2011) and the practical engineering experience.

4 Slope protection engineering is safe or not, besides the direct correlation with the structural arrangement of the surface itself, is closely related to the protection of the foot or the bottom of the dike. To take protective measures properly is one of the key problems related to the success or failure of the wave prevention project. The bottom effect is to prevent the foundation soil in front of the dike from being washed, which causes the fall or partial collapse of the laminated riprap, finally affecting the stability of the embankment. The thickness and width of the rubble layer and the thickness of the gravel layer under the bottom of the retaining block are determined according to the engineering experience and the *Design and Construction Regulations of the Breakwater* (JTS 154-1—2011).

7.18 Substructure in reservoir area

7.18.1 The substructure in reservoir area is the substructure which is built along the side of the reservoir or across its tributary channel and tributary and affected by the water level in the reservoir. Because of the damage factors, such as reservoir water level fluctuation, wave attack, ground water mound, reservoir bank ruin, water erosion, reservoir siltation and others, to reservoir banks and substructure, design need to be based on these adverse factors, distinguishing between primary and secondary, and take appropriate measures of protection and reinforcement.

7.18.2 Under the action of seepage pressure, it will reduce the stability of the embankment side slope, may result in piping and soil flow phenomenon, and is not conducive to the stability of the embankment. Therefore, it is necessary to use well graded permeable material as filling material, and strictly control the degree of compaction of embankment filling.

7.18.3 For the embankment across the branch ditch, when the water level in the ditch is much higher than the reservoir water level, it will produce a stable seepage in substructure; if the water level difference between the upper and lower reaches is not obvious, the sudden change of water level will lead to unstable seepage in the substructure when the reservoir is released by flood or flood is coming. The amplitude and time change of reservoir water level is complex. It is generally believed that when the soil permeability coefficient $K > 0.001$ cm/s, the water level extinction velocity is less than 1 m/d, and larger than 3 m/d is a sudden drop. When the reservoir water level drops suddenly, the water on the inside of the embankment flows to the reservoir area, and the seepage pressure and erosion on the reservoir side slope are produced. When the water level rises, the water in the reservoir flows to the embankment and it will produce the seepage pressure and erosion on the outside side slope.

Most of the embankment instability occurs when the water level falls suddenly. The strip method with assumed rupture surface as circular surface is generally used in checking calculation, but the change of body weight and strength index must be counted.

In the fast silting reservoir area, because the initial stage of water storage is a dangerous period, there is no deposit or less deposit, when checking the stability of embankment, the sliding resistance of embankment will not be considered after the increasing deposition in the future.

The seepage deformation in embankment is mainly piping and flowing soil. The design of reservoir substructure needs to be analyzed the permeation failure from the aspects of soil heterogeneity coefficient, particle diameter, soil compactness and permeability coefficient, seepage velocity, seepage pressure and so on.

After impoundment, with the rise and fall of water level, the ground water rises, the dynamic action of waves and the change of reservoir formation after immersion, it destroys the stability of the existing slope, makes the reservoir shore erosion, collapse, landslide and other deformation. During the design, the stability of the reservoir bank should be evaluated according to the specific location of the highway. When endangering highway safety, the bank or embankment needs to be protected and reinforced.

7.18.4 The slope surface of the embankment immersed in the reservoir area is mainly to prevent the wave invasion. In the upstream area of reservoir, because of the increase of the flow velocity, the scouring effect of the water flow must be considered.

Setting stronger protection engineering in the protection scope of domestic demand, to resist the impact of wave damage. The immersed waterway embankment is paved with dry masonry stone or various types of concrete blocks (slabs), and the reverse filter layer needs to be done between the protective building and the soil to prevent the loss of soil particles.

7.18.6 After impoundment, changes of physical and mechanical properties of strata under reservoir bank and the scour effect of waves on bank will produce the deformation phenomena of sliding and collapsing.

The speed of the collapse of the reservoir varies depending on the soil bank and its location. According to the comprehensive consideration of the distance from the highway to the shore of the reservoir and the other above factors, it is necessary to distinguish priorities, and achieve staging investment. For the section where have heavy collapse tendency and the recent threat to safety of substructure, it needs to do the bank protection project, when the new construction.

Table 7.11 Division of frozen zone table

Division of frozen zone	Heavy frozen area	Middle frozen zone	Light frozen area	Non freezing zone
freezing index (°C)	≥2000	2000–800	800–50	≤50

7.19 Substructure in seasonally frozen soil area

7.19.1 The main cause of highway distress in substructure in seasonally frozen soil area is the frost heave and thaw collapse. The main factors affecting frost heave is ground water, soil texture and temperature.

According to the relevant researches, the frost-heave capacity of substructure makes up around 90% of the total frost heave of highway, which means that the frost-heave capacity of substructure can represent the total frost heave of highway. Subgrade in seasonally frozen soil area in China is extensive in scope. The geological conditions of Qinghai-Tibetan Plateau, northwest China and northeast China are significantly different, including plateau, desert, Gobi, meadow moor, mountains and plains. Even in the same area, the difference in the substructure filling, height, groundwater level and the set of the side ditches also attribute to the difference in frost resisting of the substructure. For this reason, the designers need to make the design of freeze–thaw resistance for the substructure according to the specific circumstance, geological conditions. It also needs to combine wither local engineering experience.

7.19.2 *Specifications for Design of Highway Asphalt Pavement* (JTG D50—2006)divides the freeze zone according to the freezing index shown in Table 7.11. Combined with relevant standards and the actual degree of freeze–thaw damage, the seasonally frozen soil areas in this specification are mainly referred to the moderate and severe areas which the freezing index is over 800.

There exists a slight difference in the classification of frozen heaving of soil in different industries and specifications. The Table 7.12 shows the classification of frozen heaving of soil in different specifications. The classifications mainly refer to specifications for construction departments in China.

7.19.4 The substructure fillings are important in reducing the frost heaving. The different coefficients of frost heaving have different influence, especially on the intensity difference after substructure melted. Coarse aggregate will still keep high intensity and meet the demand for the road after melting even forms the frost heaving. So choosing a good filling is the fundamental condition to guarantee the quality of the substructure.

The excavation section is an artificial lowland and the ground water in higher place will percolate into the through cut. For this reason, the frost heaving problems for the through cut is more serious than the fill soil. The requirement of soil texture for the slope section is higher in order to control frost heave and reduce frost damage effectively.

Vaporous water, surface water seepage and capillary water are included in the categories of frost heaving water of soil. The frost heave for gravel material is mainly

Table 7.12 Classification of frost heave degree of different specifications soil

Frost heave classification	Name of code		
	Technical guidelines for frost resistant design and construction of highway engineering and code for building foundations in permafrost regions	Code for design of ground base and foundation of highway bridges and culverts	Code for design of flexible pavement of traffic construction ministry of the former Soviet Union (1985)
Non frost heave	$\eta \leq 1.0$	$\eta \leq 1.0$	$\eta \leq 1.0$
Weak frost heave	$1 < \eta \leq 3.5$	$1 < \eta \leq 3.5$	$1 < \eta \leq 4.0$
Frost heave	$3.5 < \eta \leq 6.0$	$3.5 < \eta \leq 6.0$	$4.0 < \eta \leq 7.0$
Strong frost heave	$6.0 < \eta \leq 12.0$	$6.0 < \eta \leq 13.0$	$7.0 < \eta \leq 10.0$
Extra strong frost heave	$\eta > 12.0$	$\eta > 13.0$	$\eta > 10$

about the condensation of gaseous water, which is similar to the frozen on the window panes in winter. Although using gravel material can block the capillary water, it cannot avoid frozen completely. This conclusion is convinced in the survey of several highways. The substructure in Nordic Countries face the problem of low height and low temperature. The government pays attention to the handle of the substructure and mainly uses gravel and other coarse materials in filling. The problem of frost heave is not serious then.

The damage of frost heave usually occurs in the spring thawing period. The spring thawing causes rate of water content increasing and the substructure strength dropping sharply. Under the live road of automobile, the damages like crack, frost heave, sinkage, tracing rut and upheaval will appear. So it is essential to guarantee the intensity of the substructure during the period of spring thawing. The water permeability of gravel material is good, which can rapidly eliminate the thawing water and keep a high level of intensity even with a high rate of water content. Curing agent like cement, lime and fly ash can be used to stabilize granule soil in areas where gravel materials are scarce. According to the analysis of five years of observation in a test road in Heilongjiang Province, the strength retrogression coefficient for the material of underlayer and substructure under the repeated action of freezing and thawing is following: cement stabilized gravel 20–25%, calcareous soil 30–40%, sand bedding course 25–30%, substructure 25–30%, and the stabilized fine grained soil has relatively poor long term intensity after freezing and thawing.

Additional References for English Version

[1] Pavement structure consists of surface course, base course, subbase course (optional) and subgrade (Fig. 1). Types of road cross section include filler, cut, half filler and half cut (Fig. 2). The word 'substructure' will represent three types of road cross section including pavement structure. The components of substructure are shown in Fig. 3.

[2] 'Width of substructure' represents the whole width from right-of-way line to right-of-way line which includes travelled way, shoulder and median (optional) (Fig. 3).

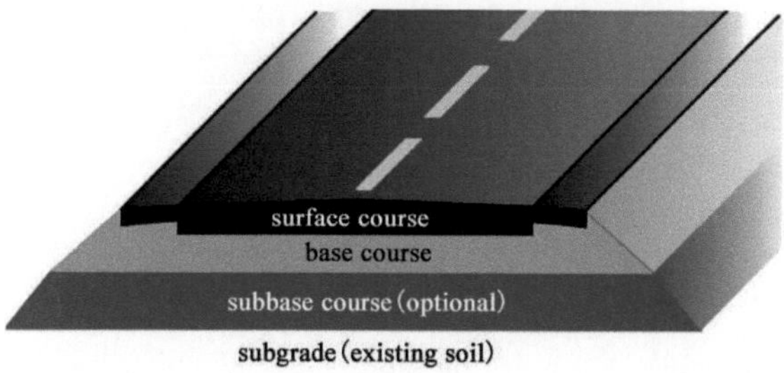

Fig. 1 Pavement structure

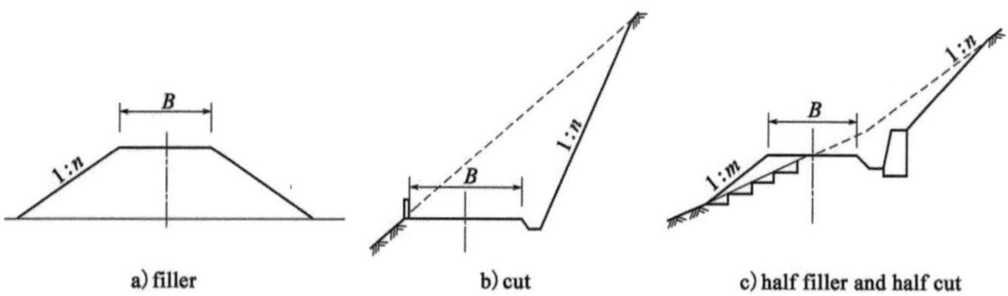

Fig. 2 Road cross section

W. Wu, *Specifications for Design of Highway Substructure*, JTG Standards and Specifications, https://doi.org/10.1007/978-981-96-2313-6

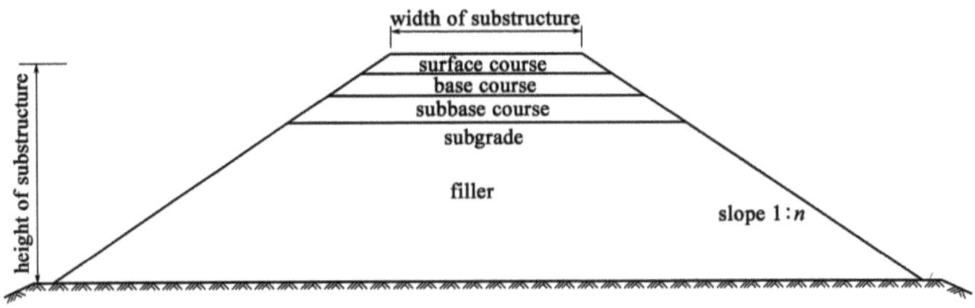

Fig. 3 Components of substructure

[3] 'Height of substructure' represents the height from designed elevation to the ground elevation (Fig. 3).

MIX
Papier aus verantwortungsvollen Quellen
Paper from responsible sources
FSC® C105338

If you have any concerns about our products,
you can contact us on
ProductSafety@springernature.com

In case Publisher is established outside the EU,
the EU authorized representative is:
Springer Nature Customer Service Center GmbH
Europaplatz 3, 69115 Heidelberg, Germany

Printed by Libri Plureos GmbH
in Hamburg, Germany